Weston, Ralph E.
Chemical kinetics

Chemical Kinetics

FUNDAMENTAL TOPICS IN PHYSICAL CHEMISTRY

Editor: **Harold S. Johnston,** *University of California, Berkeley*

HANNAY   *Solid-state Chemistry*

SOMORJAI   *Principles of Surface Chemistry*

STRAUSS   *Quantum Mechanics: An Introduction*

WESTON and SCHWARZ   *Chemical Kinetics*

RALPH E. WESTON, JR. and HAROLD A. SCHWARZ

Brookhaven National Laboratory

Chemical

Kinetics

PRENTICE-HALL, INC.

Englewood Cliffs, New Jersey

FUNDAMENTAL TOPICS IN PHYSICAL CHEMISTRY
Editor, Harold S. Johnston

Printed in the United States of America.

Library of Congress Catalog Card No.: 79-165559

13-128660-9

Current printing
10
9
8
7
6
5
4
3
2

PRENTICE-HALL INTERNATIONAL, INC., *London*
PRENTICE-HALL OF AUSTRALIA, PTY. LTD., *Sydney*
PRENTICE-HALL OF CANADA, LTD., *Toronto*
PRENTICE-HALL OF INDIA PRIVATE LIMITED, *New Delhi*
PRENTICE-HALL OF JAPAN, INC., *Tokyo*

Foreword

has shifted over the past generation from the macroscopic view of thermodynamics, electrochemistry, and empirical kinetics to the molecular view of these and other areas. Statistical mechanics has provided a practical method for the evaluation of certain types of thermodynamic data. The microscopic viewpoint of solid-state physics has injected new intellectual vigor into electrochemistry, surface chemistry, and heterogeneous catalysis. Recent developments in molecular beams have turned chemical kinetics into one of the most active fields of physical chemistry. Thus most areas of physical chemistry now present two aspects, the old and the new; and this dichotomy is reflected in the teaching of physical chemistry.

The usual curriculum in chemistry includes a year's course in physical chemistry given during the sophomore or junior year. In some cases, teachers of this course try to retain all the old physical chemistry and to introduce all the new topics as well; then the sheer volume of material is overwhelming. In other cases, an effort is made to cover all the material from the modern point of view, starting with quantum mechanics; this approach tends to neglect (or treat in only a superficial way) complicated molecules, solutions, and the liquid state. It ought to be recognized that physical chemistry is not a body of knowledge that must be covered but rather a set of methods for predicting chemical events. Each of the methods should be understood. It is desirable to present chemical thermodynamics by the macroscopic approach, lest this method of science be omitted from the student's education. Next, it is imperative to teach the student a small amount of rigorous, Schrödinger-type quantum mechanics, with some—but not all—applications to molecular structure and molecular spectroscopy. The student of physical chemistry must also receive an introduction to statistical mechanics but not a survey of all its applications. With a rigorous, though limited, introduction to thermodynamics, quantum mechanics, and statistical mechanics, the student can readily be introduced to the methods and viewpoints of chemical kinetics, solid-state chemistry, and other areas of physical chemistry.

The ideal textbook in physical chemistry might contain a condensed, clear exposition of each *method of prediction* in chemistry, with extra reading material and references. Such a book would be long, and a course based on it inflexible. It therefore seems desirable to have available a series of relatively brief and inexpensive texts that can help serve the same purpose and that can give the teacher greater flexibility in his choice of teaching materials.

This then is the aim of the series Fundamental Topics in Physical Chemistry. Each author has been urged to make his goal student insight into one basic area in physical chemistry.

Harold S. Johnston

Preface

THE FIRST QUANTITATIVE STUDY of reaction rates was made a little more than one hundred years ago, but chemical kinetics has matured only within the present century. During this period it has become an indispensible tool in all fields of chemistry: physical, inorganic, analytical, organic, and biological. In addition to the contributions made by kinetics to an understanding of fundamental chemical processes, it has technological applications ranging from the fermentation of grapes to the prevention of air pollution.

A book as brief as this one cannot cover all the possible topics of chemical kinetics without treating some of them superficially. We have, therefore, omitted some of the topics often found in similar texts. The space thus gained enables us to treat the remaining topics in greater detail and with more rigor. The emphasis throughout is on the factors that determine the rates of elementary reactions; for this reason, relatively little space is devoted to the use of kinetics to determine reaction mechanisms, or to complex reactions. These are important aspects of chemical kinetics, but with this book as a background, the student should be capable of further independent reading. Our approach to kinetics begins at the molecular level, whenever this is feasible. Thus, the classical mechanics of two-body collisions is discussed extensively and is then related to molecular-beam experiments and to a theoretical analog, classical trajectory calculations. Similarly, the chapter on reactions in solution emphasizes diffusion processes and the statistical mechanics of ionic solutions.

This book is aimed at seniors and first-year graduate students who have had a background in thermodynamics, a little quantum mechanics, and the elements of statistical mechanics of equilibrium systems. A brief review of the latter subject is included in an appendix. A number of problems are included in order to give the student some skill in the arithmetic manipulations typically involved in the practice of kinetics. In many cases, the problems round out the treatment of specific topics which have been dealt with summarily in the text.

We hope that the brief introduction to a complex and rapidly growing field provided by this textbook will give the student an understanding of the basic physical principles involved in the kinetics of chemical reactions, as well as an appreciation of the new and exciting developments that are providing the subject with its present momentum.

Our thanks are due to Professor Harold S. Johnston for his generous advice and encouragement, and to him and Professor R. A. Marcus for critical reviews of the completed manuscript. One of us (R.E.W.) had many stimulating discussions of kinetics with Professor Bruce H. Mahan. Some of this text was prepared while we were guests at the Department of Chemistry, University of California at Berkeley (R.E.W.) and at the Department of Physical Chemistry, Hebrew University, in Jerusalem (H.A.S.). We acknowledge with pleasure the hospitality extended by members of those departments.

Ralph E. Weston
Harold A. Schwarz

Contents

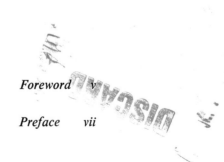

5. UNIMOLECULAR REACTIONS, 117

6. REACTIONS IN SOLUTION, 151

Chemical Kinetics

1

Parameters of

Chemical Kinetics

CHEMICAL KINETICS may be broadly defined as the study of systems whose chemical composition or energy distribution is changing with time. Thus it forms an important branch of physical chemistry that is complementary to thermodynamics, the study of systems at equilibrium. Although the principles of thermodynamics allow us to predict whether a particular chemical reaction can or cannot take place, they tell us nothing about the speed with which the reaction will occur. Thus nitrogen and oxygen in the earth's atmosphere could react with water in the oceans to form dilute nitric acid according to the equation

$$2\,H_2O + 2\,N_2 + 5\,O_2 \longrightarrow 4\,HNO_3(0.1\ M),\ \Delta G^0_{298} = -85\ \text{kcal/mole}$$

$$(1.1)$$

The negative free energy indicates that this reaction can proceed spontaneously, but the fortunate fact is that the rate is immeasurably slow.

Other important differences between thermodynamics and kinetics exist. According to fundamental laws of the former discipline, the value of an equilibrium constant for a reaction such as (1.1) is independent of the path leading from reactants to products. This fact is definitely not true of the rate of a reaction, where the individual steps along the way are extremely important. In fact, the determination of all these individual steps (which combine to form the *mechanism of reaction*) is an important part of chemical kinetics. Among the differences between thermodynamics and kinetics is one that theoretical kineticists would like to erase. Starting from molecular properties of reactants and products, one can calculate the equilibrium constant for a chemical reaction, using the methods of statistical mechanics. It is not yet possible to obtain rate constants in a rigorous way from the same kind of information.

1.1. Reaction Mechanisms

The stoichiometry of the reaction of hydrogen and bromine to form hydrogen bromide is given by

$$H_2 + Br_2 \longrightarrow 2\,HBr \tag{1.2}$$

The reaction does not occur by the collision of a hydrogen molecule with a bromine molecule, however. Instead, five separate reactions are taking place:

$$
\begin{aligned}
&Br_2 + M \text{ (any other molecule)} \longrightarrow 2\,Br + M \\
&Br + H_2 \longrightarrow HBr + H \\
&H + Br_2 \longrightarrow HBr + Br \\
&H + HBr \longrightarrow H_2 + Br \\
&2\,Br + M \longrightarrow Br_2 + M
\end{aligned}
\tag{1.3}
$$

The collection of reactions (1.3) is the *mechanism* of the reaction. Each step in (1.3) is an *elementary* reaction. Reaction (1.2) is a *complex* reaction, for more than one elementary reaction is involved in the mechanism. The actual reaction system is composed not only of bromine, hydrogen, and hydrogen bromide but also of hydrogen atoms and bromine atoms that are *intermediates* in the reaction.

One of the important goals of chemical kinetics is to discover which elementary reactions constitute the mechanism of a complex reaction. Because the dependence of the reaction rate on concentrations of reactants and products is specified by the mechanism (Section 1.3), kinetic studies are very helpful in achieving this goal. They should not be relied upon exclusively, however, and many other techniques, such as isotopic labeling, various kinds of spectroscopy, and studies of analogous reactions, are available to aid in mechanism studies. We shall say little about techniques other than reaction rate studies, in this book.

Elementary reactions can be combined in different ways to produce other complex reactions, and the information obtained about an elementary reaction in one system can be transferred to another. Thus the reaction rate obtained for the reaction of Br_2 with H in the foregoing system can also be used in the system

$H_2 \longrightarrow 2 H$ (in an electrical discharge)

$Br_2 + H \longrightarrow HBr + Br$ (1.4)

$2 Br + M \longrightarrow Br_2 + M$

This emphasizes the importance of isolating the elementary steps and determining their rates.

An important point to be made about elementary chemical reactions is that the reactants are in thermal equilibrium with their surroundings; that is, their energy states (translational, rotational, vibrational, and electronic) are determined solely by the temperature (through the laws of statistical mechanics). If such is *not* the case, we designate the reactions as *nonthermal elementary reactions.*

The photolysis of hydrogen iodide by ultraviolet light (1849 Å) produces the reaction

$HI + h\nu \longrightarrow I + H^*$

where H* is a hydrogen atom with 84 kcal/mole of excess translational energy. This energy may be dissipated by collision with inert molecules, but nonthermal reactions can also occur; for example, abstraction of a deuterium atom from either deuterium or a deuterated hydrocarbon:

$H^* + RD \longrightarrow HD + R\cdot$

This reaction can also be studied as a normal elementary reaction, but the reaction rate is not the same as that for the nonthermal case. Nonthermal reactions are also studied by the molecular beam technique, in which an energy-selected beam of one reactant crosses a beam of another reactant, and the angular distribution of the reaction products is measured relative to the direction of the beams (Appendix B).

1.2. The Rate of Reaction

Experimental studies of reactions generally involve the determination of reaction rates as a function of several variables: chemical composition, temperature, pressure, or volume. At this point it is helpful to introduce some definitions. Consider the reaction of hydrogen peroxide with ferrous iron, known as Fenton's reaction:

$$2\,Fe(II) + 2\,H^+ + H_2O_2 \longrightarrow 2\,Fe(III) + 2\,H_2O \qquad (1.5)$$

Since two moles of ferrous iron are oxidized for each mole of hydrogen peroxide that reacts,

$$\frac{d[Fe(II)]}{dt} = 2\left[\frac{d(H_2O_2)}{dt}\right]$$

It is convenient to define a single rate for the reaction as

$$\text{Rate} = \mathscr{R} = -\frac{d(H_2O_2)}{dt} = \frac{1}{2}\frac{-d[Fe(II)]}{dt}$$

In general, if the stoichiometry of a chemical reaction can be described completely by the equation

$$aA + bB + \cdots \longrightarrow pP + qQ \qquad (1.6)$$

the rate of reaction is defined by

$$\mathscr{R} = -\frac{1}{a}\frac{d(A)}{dt} = -\frac{1}{b}\frac{d(B)}{dt} = \frac{1}{p}\frac{d(P)}{dt} = \frac{1}{q}\frac{d(Q)}{dt} \qquad (1.7)$$

where the parentheses indicate concentrations and t is the time. Not all authors include the stoichiometric factors a, b, etc., in such a definition, so caution is always required in using rates quoted in the literature. The quantity $\mathscr{R}$ has units of (concentration)/(time), and we shall use units of moles liter^{-1} second^{-1}.

According to Eq. (1.7), concentrations of reactants and products are related by the stoichiometry of the reaction unequivocally; therefore the measurement of the rate of change of any one of the concentrations determines $\mathscr{R}$. This fact is rigorously true only for elementary reactions. If Eq. (1.6) represents the overall stoichiometry of a complex reaction, then reaction intermediates will build up in the early stages and disappear at the end of the reaction; thus the stoichiometry will change with time. If the reaction intermediates are in sufficiently low concentration, their effect on the stoichiometry will be negligible and Eq. (1.7) will appear valid. Fenton's reaction (1.5) is a complex reaction, as may be seen from the fact that organic compounds are oxidized when present during the reaction but not by the reactants or products separately. Obviously some intermediate is present that oxidizes the organic compounds, but the concentration of this intermediate is negligible in comparison with that of reactants or products.

The number of methods that have been used to measure reaction rates is myriad, and it is beyond the scope of this book to describe them (see General References 2, 4, 6). They may be continuous methods that

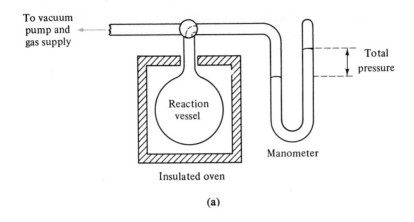

(a)

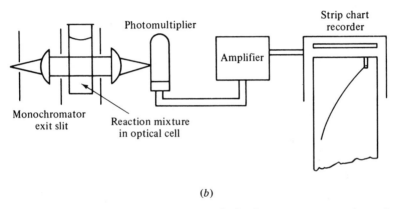

(b)

Figure 1.1. *Examples of continuous methods of rate measurement in static systems: (a) measurement of total pressure in a gaseous reaction; (b) spectrophotometric measurement of reactant or product concentration in a liquid-phase reaction.*

enable the concentration to be determined without disturbing the reaction system, such as spectrophotometry (visible, UV, or IR), electrical conductivity, and pressure (in gaseous reactions). In other instances (although much less frequently than in the past) it is necessary to interrupt the reaction and to analyze the entire reaction mixture. Another alternative is to withdraw a small sample from the mixture, "quench" one of the reactants, and save another reactant or product for subsequent analysis. In these cases, standard procedures of quantitative analysis are used: acid-base titrimetry, colorimetric methods, chromatography, including gas-liquid chromatography, and so on. Ideally one should use a method that is very specific (not the total pressure, for example), continuous, rapid, and incapable of disturbing the reaction system. Ideal conditions are rarely found.

Not only can one use different methods of analysis, but one can also study the reaction in a static system (for example, those of Fig. 1.1) or in a flow system (Appendix A). The former is more conventional, whereas the latter is required to measure rates of very rapid reactions.

1.3. Concentration Dependence
of Reaction Rates

In some complex reactions and in all elementary reactions, the concentration dependence of the reaction rate takes the form

$$\mathscr{R} = k(A)^\alpha (B)^\beta \cdots \tag{1.8}$$

where (A), (B), ..., are concentrations of reactants or catalysts and where k is called the *rate constant*. The exponents α and β (which may not be integers in a complex reaction) are called the *order of reaction* with respect to A, with respect to B, and so on. The *total order* of the reaction is given by the sum of these exponents:

$$n = \alpha + \beta + \cdots \tag{1.9}$$

For example, the reaction

$$2\,NO + O_2 \longrightarrow 2\,NO_2$$

follows the rate law

$$\mathscr{R} = \frac{d(NO_2)}{2dt} = k(NO)^2(O_2)$$

and is second order with respect to nitric oxide, first order with respect to oxygen, and third order overall (or the total order is three).

The *molecularity* of an elementary reaction is determined by the number of molecules of reactants involved when the reaction is interpreted at the molecular level. If a single molecule is involved, the reaction is *unimolecular*. If two molecules are involved (either two molecules of the same reactant or two different reactants), the reaction is *bimolecular*; and if three molecules are involved, the reaction is *termolecular*. (No example is known of higher molecularities.) Molecularity has meaning only for elementary reactions, and the order and molecularity usually coincide. The only exception is for unimolecular reactions in the gas phase, which is such an important exception that a separate chapter will be devoted to this case.

One can see from Eq. (1.8) that the rate constant will have units of $(\text{concentration})^{1-n}/(\text{time})$; we shall always use $(\text{moles/liter})^{1-n}$ seconds^{-1}. A first-order rate constant will be expressed in seconds^{-1}, a second-order constant in liters mole^{-1} sec^{-1}, and so forth. Other concentration units are frequently employed, particularly for gas-phase reactions, where molecules cc^{-1} or moles cc^{-1} are common.

One should also note that some complex reactions have a concentration dependence more complicated that that of Eq. (1.8) and, in particular, that the rate may depend on product concentrations.

1.4. Integrated Forms of Rate Expressions

One possible method of determining a rate constant is based on Eq. (1.8). The concentration of one compound is measured at such small time intervals that a time derivative, say $d(\text{A})/dt$, can be approximated by $\Delta(\text{A})/\Delta t$. If the initial concentrations of the reactants are varied systematically, it then becomes possible to determine the order of reaction with respect to the various participants, and hence to calculate k.

Usually it is more convenient to have the rate expression in an integrated form that relates concentration and time. The simplest case is that of a first-order reaction. The differential rate expression is

$$\frac{-d(\text{A})}{dt} = k(\text{A}) \tag{1.10}$$

which is a standard form, yielding on integration

$$\ln \frac{(\text{A})}{(\text{A})_0} = -kt \tag{1.11}$$

If it is possible to measure (A) as a function of time, the rate constant k may be found as the slope of the line obtained when $\ln (\text{A})$ is plotted against time. It is not necessary to know the initial concentration $(\text{A})_0$; nor is it really necessary to know the absolute value of (A). Any quantity proportional to (A) (such as the absorbance of light by A, or the pressure of A) will have a similar time dependence. Suppose that a measured quantity X is related to (A) by

$$X = a(\text{A})$$

Then Eq. (1.11) becomes

$$\ln X - \ln a - \ln X_0 + \ln a = -kt$$

Again, k may be found as the slope of a plot of $\ln X$ versus t. An example is shown in Fig. 1.2. The concentrations of A and the reaction product

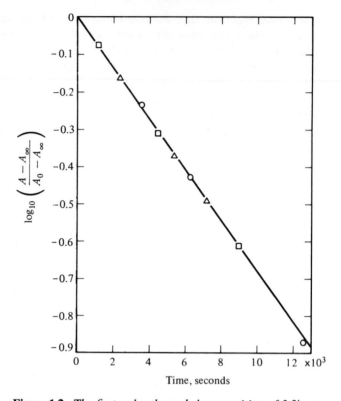

Figure 1.2. *The first-order thermal decomposition of 2,2'-azo-bis-isobutyronitrile in toluene at 80.4°C. The absorbance of the solution at the start is A_0, that at time t is A, and that after 24 hours is A_∞. Hence, the quantity $A - A_\infty$ is proportional to reactant concentration. The absorption spectrum of the nitrile has a peak at 345 nm, while the data shown were obtained at 360 nm ($\bigcirc$), 370 nm ($\square$), and 380 nm ($\triangle$). The three sets of data are normalized by use of the ratio $(A - A_\infty)/(A_0 - A_\infty)$. The rate constant found from this plot is 1.55×10^{-4} sec^{-1}. [From M. Talat-Erben and S. Bywater, J. Am. Chem. Soc. 77, 3712 (1955).]*

P are related by

$$(A)_0 - (A) = (P) - (P)_0$$

and if the reaction is studied by following (P),

$$\ln \frac{(P)_0 + (A)_0 - (P)}{(A)_0} = -kt \qquad (1.12)$$

Knowing $(P)_0$ and $(A)_0$, one can easily obtain k. It is not neccessary to know them, however, because the value of (P) after a very long time, $(P)_\infty$, is equal to $(P)_0 + (A)_0$, so that

$$\ln \frac{(P)_\infty - (P)}{(A)_0} = -kt$$

If the first-order reaction is reversible—that is,

$$A \underset{k_r}{\overset{k_f}{\rightleftharpoons}} P \tag{1.13}$$

the differential rate expression will be

$$\frac{-d(A)}{dt} = \mathscr{R}_f - \mathscr{R}_r \tag{1.14}$$

where $\mathscr{R}_f$ is the rate of the forward reaction and $\mathscr{R}_r$ is the rate of the reverse reaction. Equation (1.14) is a simple example of a general rule in kinetics: The rates for all elementary reactions in which a specific component is a reactant or product may be combined to give the net rate of concentration change for that component. One must be careful to introduce the correct sign (positive or negative) and stoichiometric factor for each elementary reaction.

If the reverse reaction is also first order, then

$$\frac{-d(A)}{dt} = \frac{d(P)}{dt} = k_f(A) - k_r(P) \tag{1.15}$$

and if $(P)_0 = 0$, $(A)_0 - (A) = (P)$.
Hence

$$\frac{-d(A)}{dt} = k_f(A) - k_r[(A)_0 - (A)]$$
$$= (k_f + k_r)(A) - k_r(A)_0$$

Integrating, we have

$$\ln \frac{k_f(A)_0}{(k_f + k_r)(A) - k_r(A)_0} = (k_f + k_r)t \tag{1.16}$$

This result can be expressed differently if we note that at equilibrium $d(A)_e/dt = 0$, and hence

$$k_f(A)_e = k_r(P)_e = k_r[(A)_0 - (A)_e] \tag{1.17}$$

Therefore

$$\frac{(A)_0 - (A)_e}{(A) - (A)_e} = \frac{(A)_0(k_f + k_r) - k_r(A)_0}{(A)(k_f + k_r) - k_r(A)_0}$$

This equation can be combined with Eq. (1.16) to give

$$\ln \frac{(A)_0 - (A)_e}{(A) - (A)_e} = (k_f + k_r)t \tag{1.18}$$

Equation (1.18) shows that the approach to equilibrium is a first-order process [see Eq. (1.11)] with a rate constant equal to the *sum* of those for the forward and reverse reactions.

There are two types of second-order reactions. In the first, the reaction is second order with respect to a single component. Hence

$$\frac{-d(A)}{dt} = 2k(A)^2 \tag{1.19}$$

which is again a standard form, yielding upon integration

$$\frac{1}{(A)} - \frac{1}{(A)_0} = 2kt \tag{1.20}$$

Again, it is not necessary to know the initial concentration $(A)_0$, but it is necessary to know the actual concentration of A in order to obtain a rate constant, a fact that might be anticipated, since the units of k involve concentration. The value of the rate constant is generally obtained as the slope of the line obtained by plotting the reciprocal of (A) versus time.

The more usual type of second-order reaction is first order in each of two different reactants. The rate law is

$$\frac{-d(A)}{dt} = k(A)(B) \tag{1.21}$$

If the initial concentrations of A and B are equal, the integration follows that just discussed. More generally, if one mole of A reacts with one mole of B,

$$(A)_0 - (A) = (B)_0 - (B)$$
$$(B) = (A) - (A)_0 - (B)_0$$

and we have

$$\frac{-d(A)}{dt} = k(A)[(A) - (A)_0 + (B)_0]$$

This equation may be integrated by the method of partial fractions. The result is

$$\frac{1}{(B)_0 - (A)_0} \ln \frac{(A)_0}{(A)} - \frac{1}{(B)_0 - (A)_0} \ln \frac{(B)_0 - (A)_0 - (A)}{(B)_0} = kt$$

which may be arranged into the more convenient form

$$\frac{1}{(B)_0 - (A)_0} \ln \frac{(A)_0(B)}{(B)_0(A)} = kt \tag{1.22}$$

It is quite usual to arrange the experiment so that one of the reactants, say B, is present in very large excess. In this case,

$$(B) \cong (B)_0 \gg (A)_0$$

and Eq. (1.22) reduces to

$$\ln \frac{(A)_0}{(A)} = k(B)_0 t \qquad (1.23)$$

This equation will also hold rigorously for reactions in which B is a catalyst—that is, a substance whose concentration does not change during the course of the reaction.

Note the resemblance between Eq. (1.23) and Eq. (1.11) for a first-order reaction. Systems that follow Eq. (1.23) are referred to as *pseudo-first-order* reactions. The apparent first-order rate constant is converted into the second-order constant, k, upon division by the initial concentration $(B)_0$. Pseudo-first-order reactions are very useful for studying the kinetics of short-lived reaction intermediates, since it is not necessary to measure the actual concentration of the intermediate but only some property proportional to it.

1.5. The Principle of Detailed Balancing

The introduction of reversible reactions in Section 1.4 raises the question of the relationship between reaction rates and equilibria. The *principle of detailed balancing** asserts that at equilibrium the rates of the forward and reverse processes are equal for every elementary reaction occurring.

For the specific example of a reversible first-order reaction,

$$k_f(A)_e = k_r(P)_e$$

and

$$\frac{(P)_e}{(A)_e} = \frac{k_f}{k_r} = K$$

where K is the usual chemical equilibrium constant. Thus it becomes possible to evaluate both k_f and k_r if the rate of approach to equilibrium as given in Eq. (1.18) and the equilibrium constant are both measured.

---

*A more fundamental concept is that of *microscopic reversibility*, which states that at equilibrium the rate of transition from microscopic state i to state j is identical with that for the reverse transition $(j \longrightarrow i)$. This is a consequence of the fact that the fundamental equations for the microscopic dynamics of a system (i.e., Newton's laws or the Schrödinger equation) have the same form when t is replaced by $-t$ and when the signs of all velocities are also reversed (1).

Although this particular example is rather obvious, the principle of detailed balancing is applicable to more complex situations. Consider three reversible equilibria, interconnected in the following way:

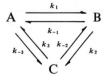

Then

$$\frac{(B)_e}{(A)_e} = \frac{k_1}{k_{-1}} = K_1 \qquad \frac{(C)_e}{(B)_e} = \frac{k_2}{k_{-2}} = K_2$$

and

$$\frac{(A)_e}{(C)_e} = \frac{k_3}{k_{-3}} = K_3$$

Combining the last two equalities, we have

$$\frac{(B)_e}{(A)_e} = \frac{k_{-2}k_{-3}}{k_2 k_3} = (K_2 K_3)^{-1}$$

Therefore

$$K_1^{-1} = K_2 K_3 \quad \text{or} \quad \frac{k_1}{k_{-1}} = \frac{k_{-2}k_{-3}}{k_2 k_3} \tag{1.24}$$

This equation is another way of expressing the thermodynamic fact that the *equilibrium* between A and B does not depend on the reaction path followed. It may also be seen that the principle of detailed balancing rules out a reaction sequence with k_1, k_2, and k_3 all finite, whereas k_{-1}, k_{-2}, and k_{-3} are zero. This would describe a nonreversible cyclic system

1.6. Half-Lives of Reactions

The time required for the concentration of a reactant to decrease to one-half of its initial value, called the half-life, is a commonly used quantity that is useful because it gives a feeling for the time scale of the experiment. The half-life may be simply related to rate constants and concentrations for first-order reactions, pseudo-first-order reactions, and reactions that are second order in one component. The following relationships may be found by inserting $(A) = \frac{1}{2}(A)_0$ in Eqs. (1.11), (1.23), and (1.20):

First order $\qquad t_{1/2} = \dfrac{\ln 2}{k} = \dfrac{0.693}{k}$ $\qquad\qquad$ (1.25)

Pseudo-first order $\quad t_{1/2} = \dfrac{\ln 2}{k(B)_0}$ $\qquad\qquad$ (1.26)

Second order $\qquad t_{1/2} = \dfrac{1}{2k(A)_0}$ $\qquad\qquad$ (1.27)

For first-order and pseudo-first-order reactions, the half-life is independent of the concentration of A. For second-order reactions, the half-life is inversely proportional to $(A)_0$, and hence the second half-life period of the reaction will be twice as long as the initial half-life period.

1.7. The Effect of Pressure and Volume on Reaction Rates

Reaction rates are normally determined with the reaction mixture held at constant volume in the case of gaseous systems or at constant pressure (usually one atmosphere) in the case of solutions. The reason is simply a matter of experimental convenience. In the former case, the partial pressures of reactants correspond to concentrations. Reaction rates and rate constants should be independent of the volume of the system; if they are not it is a sure sign that a heterogeneous process is occurring at one of the surfaces. Reaction rates of unimolecular reactions in the gas phase are dependent on the pressure (see Chapter 5), but other reaction rates and rate constants are essentially independent of pressure at normal pressures. A small but important pressure dependence is noted at very high pressures (a few thousand atmospheres), for reasons explained in Chapter 6.

1.8. Temperature Dependence of Reaction Rates

A more important variable is the temperature of the reaction system. Usually, by means of a thermostat, the rate is measured under isothermal conditions. (Occasionally, if the rate of reaction is fast compared with the rate of heat transfer from the surroundings, adiabatic conditions will prevail. This process may occur in highly exothermic or endothermic gas-phase reactions.) Empirically, the rate constant of an elementary reaction is found to have a temperature dependence given by

$$k = A \exp\left(\frac{-E_a}{RT}\right) \quad \text{or} \quad \ln k = \ln A - \left(\frac{E_a}{RT}\right) \qquad (1.28)$$

This result was first formulated in 1887 by Arrhenius (2), and it is one of the facts basic to all theories of chemical kinetics. The quantity E_a is the *activation energy*, usually quoted in calories/mole or kilocalories/mole. The quantity A is called the *frequency factor* (for first-order reactions), the *pre-exponential factor*, or simply the *A factor*, and it is in the same units as the rate constant k. The significance of these constants is discussed in later chapters.

According to Eq. (1.28), a semilogarithmic plot of rate constant against the inverse of temperature should be a straight line with slope E_a/R ($E_a/R \ln 10$, if units of $\log_{10}$ are used) and intercept $\ln A$ (Fig. 1.3). Such

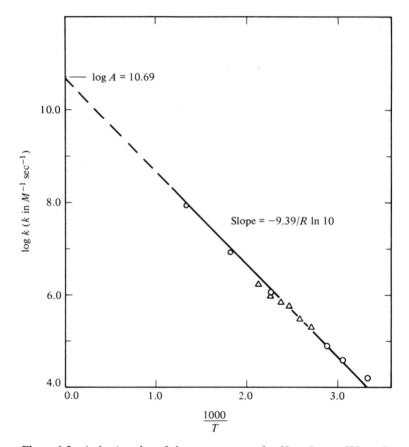

Figure 1.3. *Arrhenius plot of the rate constant for* $H + D_2 \longrightarrow HD + D$. ○, *data from A.A. Westenberg and N. de Haas,* J. Chem. Phys. **47,** *1393 (1967);* △, *data from W.R. Schulz and D. J. LeRoy,* Can. J. Chem. **42,** *2480 (1964). The experimental points at low temperatures deviate from the straight line.*

a graph is frequently referred to as an "Arrhenius plot." Over a very wide temperature range, the temperature dependence of the rate constant may not be adequately represented by this expression, and additional terms are needed. Also, the temperature dependence of the collision rate in bimolecular gas-phase reactions is sometimes explicitly included by use of the expression

$$k = A'T^{1/2} \exp\left(\frac{-E_a'}{RT}\right) \tag{1.29}$$

However, the precision of experimental data rarely justifies the use of these more complicated expressions. Really large deviations from the Arrhenius law generally indicate that the rate constant being measured is a combination of rate constants for elementary reactions with different activation energies.

1.9. Heterogeneous Reactions

Finally, we should mention a variable that is important, but one that we intend to ignore throughout this book. This is the effect of surfaces on rates of reaction—the domain of heterogeneous catalysis. All the kinetic processes included in this book are assumed to be homogeneous processes, either in the gas or liquid phase. Nevertheless, it is frequently necessary in the experimental study of reactions in the gas phase to include appropriate tests for homogeneity. These tests may consist of the introduction of additional surface in the form of glass beads, etc., or of changing the surface/volume ratio of the reaction vessel.

1.10. Competitive Reactions

Since elementary reactions can be combined in an almost limitless number of ways to form complex reactions, a wide variety of rate expressions results. A description of the methods of integrating these expressions becomes a demonstration of one's mathematical ingenuity but adds little to the understanding of kinetic processes. However, we shall give a few examples that illustrate important principles.

It is sometimes desirable to compare rate constants for a series of related compounds without determining absolute values. For example, one might want to study the solvolysis of a series of alkyl halides in order to observe the effect of various substituent groups on the reaction rate. This may be done by competitive reactions in which two or more reagents react with a common reactant:

$$A + C \xrightarrow{k_a} P$$
$$B + C \xrightarrow{k_b} Q$$

(1.30)

The reaction of A with C, having a second-order rate constant k_a, forms one set of products, while in the same solution the reaction of B with C produces another set, Q. The reactions are assumed to be elementary reactions so that the reaction orders are determined by the stoichiometry as written. The rates are

$$\frac{-d(A)}{dt} = k_a(A)(C)$$

and

$$\frac{-d(B)}{dt} = k_b(B)(C)$$

The ratio between these two rate equations yields a third equation independent of time and the concentration of C:

$$\frac{d(A)}{d(B)} = \frac{k_a(A)}{k_b(B)}$$

which may be integrated to give

$$\ln \frac{(A)}{(A)_0} = \frac{k_a}{k_b} \ln \frac{(B)}{(B)_0}$$

(1.31)

If the initial concentrations are known, the reaction may be stopped at any time and analysis for (A) and (B) will yield the ratio of rate constants, k_a/k_b, without any actual rate measurements. Since

$$(A) = (A)_0 - (P), \text{ etc.}$$

Eq. (1.31) can also be put into the form

$$\ln \left[1 - \frac{(P)}{(A)_0} \right] = \frac{k_a}{k_b} \ln \left[1 - \frac{(Q)}{(B)_0} \right]$$

and if $(P)/(A)_0$ and $(Q)/(B)_0$ are small, this is just

$$\frac{k_a}{k_b} = \frac{(P)(B)_0}{(Q)(A)_0}$$

(1.32)

Competitive reactions are often found in mechanisms, but the actual form of the integrated rate expressions will depend on the framework of the whole mechanism. The common feature of all the various forms of competition kinetics is that *relative* rate constants or rate constant ratios may be found without measuring individual rates.

1.11. Consecutive Reactions: Steady-State and Equilibrium Assumptions

Consider the consecutive first-order reaction sequence

$$A \xrightarrow{\ k_1\ } B$$
$$B \xrightarrow{\ k_2\ } C$$

(1.33)

with rate expressions

$$\frac{-d(A)}{dt} = k_1(A)$$

$$\frac{d(B)}{dt} = k_1(A) - k_2(B)$$

(1.34)

$$\frac{d(C)}{dt} = k_2(B)$$

Hence

$$(A) = (A)_0 e^{-k_1 t}$$

and

$$\frac{d(B)}{dt} = k_1(A)_0 e^{-k_1 t} - k_2(B)$$

This first-order and first-degree differential equation, linear in (B), can be solved by standard methods (3), and we find that

$$(B) = \frac{(A)_0 k_1}{k_2 - k_1}(e^{-k_1 t} - e^{-k_2 t}) \qquad \text{if } (B)_0 = 0$$

(1.35)

If, in addition, $(C)_0 = 0$, we can use the constraint

$$(A) + (B) + (C) = (A)_0$$

to solve for (C). The result is

$$(C) = (A)_0 \left[1 - e^{-k_1 t} - \frac{k_1}{k_2 - k_1}(e^{-k_1 t} - e^{-k_2 t}) \right]$$

(1.36)

The derivation of approximate rate expressions for this and other complex reactions can be greatly simplified by use of the *steady-state approximation*. This method depends on the assumption that concentrations of reaction intermediates are smaller than the concentrations of reactants or products, hence that the rate of change of intermediate concentrations is negligible compared with corresponding rates for reactants and products. Note that this assumption must be justified (usually a posteriori) in each case where it is used. Suppose that, in the preceding example, a steady-state approximation is made for (B):

$$\frac{d(B)}{dt} = 0$$

and the rate expressions (1.34) lead to

$$k_1(A)_{ss} = k_2(B)_{ss}$$

$$(B)_{ss} = \left(\frac{k_1}{k_2}\right)(A)_0 e^{-k_1 t} \tag{1.37}$$

Then

$$(C)_{ss} = (A)_0 - (A)_{ss} - (B)_{ss}$$

$$= (A)_0 \left[1 - e^{-k_1 t} - \left(\frac{k_1}{k_2}\right)e^{-k_1 t}\right]$$

$$= (A)_0(1 - e^{-k_1 t}) \qquad \text{if } (B)_{ss} \ll (A) \tag{1.38}$$

It is apparent that these approximate expressions for (B) and (C) result from the exact expressions [(1.35) and (1.36)] when the requirements

$$k_1 \ll k_2 \quad \text{and} \quad k_2 t \gg 1$$

are met. The first of these guarantees that the concentration of B will always be small, while the second requirement allows sufficient time for the buildup of product C. The results of the steady-state and exact solutions for this example are shown in Fig. 1.4.

Because of the importance of this method, it is worthwhile to apply it to a slightly more complex reaction mechanism:

$$\begin{aligned} E + S &\underset{k_{-1}}{\overset{k_1}{\rightleftharpoons}} ES \\ ES &\overset{k_2}{\longrightarrow} P + E \end{aligned} \tag{1.39}$$

The reaction intermediate, ES, is not in equilibrium with E and S even though the first reaction is reversible, because ES also reacts in an irreversible manner in the second reaction. The rate expressions are

$$\frac{d(ES)}{dt} = \mathscr{R}_1 - \mathscr{R}_{-1} - \mathscr{R}_2 \tag{1.40}$$

$$= k_1(E)(S) - k_{-1}(ES) - k_2(ES)$$

$$\frac{d(E)}{dt} = -\mathscr{R}_1 + \mathscr{R}_{-1} + \mathscr{R}_2 \tag{1.41}$$

$$\frac{d(S)}{dt} = -\mathscr{R}_1 + \mathscr{R}_{-1} \tag{1.42}$$

$$\frac{d(P)}{dt} = \mathscr{R}_2 \tag{1.43}$$

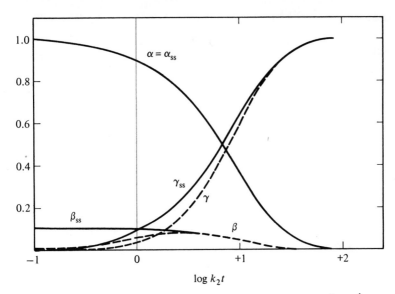

Figure 1.4. *Concentrations in the consecutive reactions* $A \xrightarrow{k_1} B \xrightarrow{k_2} C.$
The dimensionless quantities $\alpha = (A)/(A)_0$, $\beta = (B)/(A)_0$, *and* $\gamma = (C)/(A)_0$
are plotted against $\log (k_2 t)$ *for the case where* $k_2 = 10k_1$. *Solid lines are
concentrations from the steady-state approximation, and broken lines are
those from the exact rate expression.*

Suppose that a steady-state assumption is valid for (ES), so that

$$\frac{d(ES)}{dt} = 0 = \mathscr{R}_1 - \mathscr{R}_{-1} - \mathscr{R}_2 \tag{1.44}$$

It follows immediately from Eqs. (1.41) and (1.44) that

$$\frac{d(E)}{dt} = 0$$

Also, combining Eqs. (1.42) and (1.44), we have

$$\frac{-d(S)}{dt} = \mathscr{R}_2 = \frac{d(P)}{dt} = k_2(ES)_{ss} \tag{1.45}$$

From (1.44), the steady-state concentration of ES is

$$(ES)_{ss} = \frac{k_1(E)(S)}{k_{-1} + k_2}$$

and if the sum of (E) and $(ES)_{ss}$ is $(E)_0$

$$(ES)_{ss} + \frac{k_{-1} + k_2}{k_1(S)}(ES)_{ss} = (E)_0$$

$$(ES)_{ss} = \frac{(S)(E)_0}{(k_{-1} + k_2)/k_1 + (S)}$$

Inserting this equation into (1.45), we have

$$\frac{-d(S)}{dt} = \frac{k_2(S)(E)_0}{(k_{-1} + k_2)/k_1 + (S)} \tag{1.46}$$

An example of such a reaction scheme is the Michaelis-Menten mechanism (4) for enzyme reactions in which E represents the enzyme, S is the substrate, ES is a complex of the two, and P is the product.

A further simplification in the preceding rate expression is produced if, in addition to the steady-state requirements, the conditions

$$k_2 \ll k_{-1} \quad \text{and} \quad \frac{k_{-1}}{k_1} = K_s \gg (S)$$

are also met, in which case the first reaction will be very close to equilibrium and the concentration of ES will be very small compared to (E). The rate expression for (S) is now

$$\frac{-d(S)}{dt} = \frac{k_2}{K_s}(E)_0(S)$$

where K_s is the dissociation constant of ES. It is seen that the concentration of S is given by a pseudo-first-order rate expression.

Depending on the relative magnitude of the three rate constants in reaction (1.39), either the first or second elementary reaction will be *rate-determining*—that is, the overall rate of formation of product will be controlled by the rate of one of the elementary reactions. When the equilibrium assumption is applicable, this rate-controlling step is obviously the reaction of the intermediate, which is always present at its equilibrium concentration. If the steady-state assumption is valid but the ratio k_2/k_{-1} is very large (instead of very small, as for the equilibrium case), then the first step is rate-determining. As each molecule of the intermediate is formed, it rapidly reacts.

1.12. The Microscopic Approach
to Chemical Kinetics

We have discussed briefly the measurements and mathematical expressions that suffice to define the kinetics of a chemical reaction at the macroscopic level. Until fairly recently most experimental work was directed toward this goal. However, more sophisticated experiments (described in Chapter 3) enable one to obtain more detailed information about the reaction process. In order to exploit this information, and in order to develop theories of chemical kinetics based on well-established physical principles, it is necessary to move from a macroscopic to a molecular level of description. The approach we shall use in the next few chapters is as follows:

1. The starting point is a knowledge of the forces between reacting particles, which may be ions, atoms, or molecules. These may be classical electrostatic forces or forces that result from chemical bonding.

2. With this information, and hence an expression for the potential energy as a function of distances between particles, the appropriate equations of motion (classical or quantum mechanical) are solved for particles with a particular energy (translational and internal).

3. The chemical reaction process can be defined in terms of suitable trajectories of the individual particles. Thus an effective cross section for the collision process producing reaction is determined.*

4. This cross section is averaged over a distribution function that accounts for the population of molecules in each energy level or energy range. It is this final step that provides the connecting link between the microscopic and macroscopic description, as we shall see later.

References

1. N. Davidson, *Statistical Mechanics*, McGraw-Hill, New York, 1962, pp. 230 ff.
2. S. Arrhenius, *Zeitschrift für physikalische Chemie* **4**, 226 (1889).
3. F. B. Hildebrand, *Advanced Calculus for Applications*, Prentice-Hall, Englewood Cliffs, N.J., 1962 p. 7.
4. L. Michaelis and M. Menten, *Biochemische Zeitschrift* **49**, 333 (1913).

General References

1. I. Amdur and G. G. Hammes, *Chemical Kinetics: Principles and Selected Topics*, McGraw-Hill, New York, 1966.
2. S. W. Benson, *The Foundations of Chemical Kinetics*, McGraw-Hill, New York, 1960.
3. D. L. Bunker, *The International Encyclopedia of Physical Chemistry and Chemical Physics*, Topic 19, Vol. 1, "Theory of Elementary Gas Reaction Rates," Pergamon Press, Oxford, 1966.

*The concept of reaction cross section is useful in gas-phase reactions only.

4. A. A. Frost and R. G. Pearson, *Kinetics and Mechanism*, 2nd ed, John Wiley & Sons, New York, 1961.
5. H. S. Johnston, *Gas-Phase Reaction Rate Theory*, Ronald Press, New York, 1966.
6. K. J. Laidler, *Chemical Kinetics*, 2nd ed., McGraw-Hill, New York, 1965.
7. K. J. Laidler, *Theories of Chemical Reaction Rates*, McGraw-Hill, New York, 1969.

Problems

1.1. Using Eq. (1.29), derive an expression that relates the order of a reaction to reaction half-lives measured with two different initial concentrations of the reactant.

1.2. Use the steady-state approximation to set up the rate expressions for the following reactions. The intermediates whose concentrations can be considered very small are indicated in each case.

(*a*) Overall reaction: $H_2 + Br_2 \longrightarrow 2\,HBr$

Intermediates: H, Br

Mechanism:

$Br_2 + M \longrightarrow 2\,Br + M \qquad k_1$

$Br + H_2 \longrightarrow HBr + H \qquad k_2$

$H + Br_2 \longrightarrow HBr + Br \qquad k_3$

$H + HBr \longrightarrow H_2 + Br \qquad k_4$

$2\,Br + M \longrightarrow Br_2 + M \qquad k_{-1}$

(*b*) Overall reaction: $N_2O_5 \longrightarrow 2\,NO_2 + \tfrac{1}{2}O_2$

Intermediates: NO, NO_3

Mechanism:

$N_2O_5 \longrightarrow NO_2 + NO_3 \qquad\qquad k_1$

$NO_2 + NO_3 \longrightarrow N_2O_5 \qquad\qquad k_{-1}$

$NO_2 + NO_3 \longrightarrow NO + O_2 + NO_2 \qquad k_2$

$NO + NO_3 \longrightarrow 2\,NO_2 \qquad\qquad k_3$

(*c*) Overall reaction: $(CH_3)_2CO + X_2 \longrightarrow CH_2XCOCH_3 + HX$

Intermediates: $(CH_3)_2COH^+$, $CH_2C(OH)CH_3$, $CH_2XC(OH)CH_3^+$

Mechanism: (HA represents any protonic acid, X_2 a halogen)

$(CH_3)_2CO + HA \longrightarrow (CH_3)_2COH^+ + A^- \qquad\qquad k_1$

$$(CH_3)_2COH^+ + A^- \longrightarrow (CH_3)_2CO + HA \qquad k_{-1}$$

$$(CH_3)_2COH^+ + A^- \longrightarrow CH_2{=}C(OH)CH_3 + HA \qquad k_2$$

$$CH_2{=}C(OH)CH_3 + X_2 \longrightarrow CH_2XC(OH)CH_3^+ + X^- \qquad k_3$$

$$CH_2XC(OH)CH_3^+ + A^- \longrightarrow CH_2XCOCH_3 + HA \qquad k_4$$

1.3. The mechanism of Problem 1.1(c) is generally accepted as that applicable to the acid-catalyzed halogenation of acetone.

(a) From the rate expression you obtained, predict the relative rates of bromination and iodination.

(b) What is the rate-determining step if $k_2 \gg k_{-1}$?

(c) What is it if $k_2 \ll k_{-1}$?

1.4. The photolysis of hexafluoroacetone produces trifluoromethyl radicals. In the presence of added hydrogen, the radicals disappear via the reactions

$$\text{Abstraction: } CF_3 + H_2 \longrightarrow CF_3H + H \qquad k_a$$

$$\text{Recombination: } 2\,CF_3 \longrightarrow C_2F_6 \qquad k_r$$

Assuming that the recombination reaction has no energy of activation,

(a) Find E_a for the abstraction reaction, using the following data [C. L. Kibby and R. E. Weston, Jr., *Journal of Chemical Physics*, **49**, 4825 (1968)]:

1000/T (°K^{-1})	100$(k_a/k_r^{1/2})^a$
1.657	273
1.768	166
1.894	70.7
1.947	47.7
2.077	27.5
2.170	16.8
2.239	9.36
2.369	5.75
2.468	3.36
2.588	1.66
2.707	1.05
3.385	0.041

^{a}In units of $M^{-1/2}$ sec$^{-1/2}$.

(b) What is the pre-exponential factor for k_a if $k_r = 2 \times 10^{10} \, M^{-1} \, \text{sec}^{-1}$ at 400°K?

1.5. The major path for thermal decomposition of di-*t*-butyl peroxide is $(CH_3)_3COOC(CH_3)_3 \longrightarrow 2(CH_3)_2CO + C_2H_6$ [J. R. Raley, F. F. Rust, and W. E. Vaughan, *Journal of the American Chemical Society* **70**, 88 (1948)]. The rate is followed by measuring the *total pressure* (P_t) as a function of time. Use the data at the top of page 24, obtained at 147.2°C., to find the reaction order with respect to peroxide, the half-life, and the rate constant.

Time, min	P_t, Torr[a]	Time	P_t
0	179.5	26	252.5
2	187.4	30	262.1
6	198.6	34	271.3
10	210.5	38	280.2
14	221.2	40	284.9
18	231.9	42	288.9
20	237.3	46	297.1
22	242.3		

[a]Corrected for pressure of N_2 (3.1 Torr) used to force peroxide into reaction vessel.

1.6. Consider the isotopic exchange reaction

$$AX + BX^* \rightleftharpoons AX^* + BX$$

with $(AX^*) + (BX^*) \ll (AX) + (BX)$. Show that the exchange is a first-order reaction, regardless of the mechanism, and that

$$-\ln(1 - F) = \frac{\mathscr{R}[(AX) + (BX)]t}{(AX)(BX)}$$

where $F = (AX^*)_{t=t}/(AX^*)_{t=\infty}$ and $\mathscr{R}$ is the gross rate of exchange of X atoms (i.e., regardless of labeling) between AX and BX. [This expression was originally derived by H. A. C. McKay., *Nature* **142**, 997 (1938); *J. Amer. Chem. Soc.* **65**, 702 (1943).]

Suggestion: The rate at which X^* in BX^* can exchange with X in AX is proportional to the fraction $(BX^*)/[(BX^*) + (BX)]$ and to the fraction $(AX)/[(AX^*) + (AX)]$. Similarly for the reverse exchange.

1.7. The expression derived in the preceding problem is applicable to the ferrous-ferric exchange reaction (J. Silverman and R. W. Dodson, *Journal of Physical Chemistry* **56**, 846 (1952).

(*a*) Find $\mathscr{R}$ from the following data:

$(Fe(III)) = 3.11 \times 10^{-4}\ M$, $(Fe(II)) = 3.95 \times 10^{-4}\ M$, $(HClO_4) = 0.547\ M$

t, sec	0	80	330	580	890	1180	1450
F	0.00	0.07	0.34	0.43	0.57	0.67	0.74

(*b*) Find the order of reaction with respect to ferrous ion and ferric ion from the following data [$(HClO_4) = 0.547\ M$]:

$(Fe^{III}) \times 10^4\ M$	$(Fe^{II}) \times 10^4\ M$	$\mathscr{R} \times 10^8\ M\ sec^{-1}$
1.03	0.99	1.38
1.05	1.97	2.84
1.06	2.94	4.11
1.08	3.92	5.61
2.10	3.94	11.2
3.11	3.95	16.5
4.24	3.85	21.9
4.18	3.92	23.1

1.8. A graduate student has synthesized an alkyl halide, RX, which solvolyzes in water according to the equation

$$RX + H_2O \longrightarrow ROH + HX$$

He determined the rate of this reaction (assumed to be first order with respect to RX) by measuring the conductivity of the solution, which is proportional to the concentration of HX. The following data are obtained:

Time, hr	0	0.1	0.2	0.3	0.4	0.5	0.6	0.7
Λ^a	0	89.4	144.6	181.4	206.0	223.0	235.0	244.4

Time	0.8	0.9	1.0	1.2	1.4	1.6	1.8	2.0	∞
Λ	251.4	257.2	261.8	269.6	275.2	279.8	283.6	286.6	300.0

^{a} Conductivity in arbitrary units.

(*a*) Is RX a pure compound?

(*b*) If not, what can you say about the relative amounts of its components and their relative solvolysis rates?

Suggestion: Plot the data on semilogarithmic graph paper in the same way they would be plotted for a first-order reaction. How does this look when the elapsed time is large?

1.9. The rate of a gaseous reaction $A + B \longrightarrow C$ is determined by measuring total pressure. The following data are obtained at 500 °K:

P_A, Torr	P_B, Torr	Initial rate, Torr/min
50	100	0.262
100	100	1.05
100	147	1.60
100	203	2.15
153	100	2.50
198	100	4.10

(*a*) What is the reaction order with respect to A and B?

(*b*) What is the rate constant in the customary units?

2

Forces Between Atoms, Molecules, and Ions

IN THIS CHAPTER we describe briefly the types of interaction between atoms, molecules, or ions (1). For simplicity, we shall generally refer to "molecules" when the description is applicable to all three species. Although one often speaks of an interaction *force* between molecules, $F(r)$, it is usually more convenient to work with the corresponding *potential*, $V(r)$. These two quantities are related by the expression

$$F(r) = \frac{-dV(r)}{dr} \tag{2.1}$$

where r is the intermolecular distance. The potential $V(r)$ may also depend on the relative orientation of molecules, as well as on their distance apart, but we shall usually deal with cases where this effect may be neglected. Molecular interaction potentials can be divided, albeit somewhat arbitrarily, into long-range and short-range potentials.

2.1. Long-Range Forces

All long-range potentials vary inversely as powers of the intermolecular distance. The first type to be considered is the classical electrostatic or coulombic potential, which for two charges z_1e and z_2e separated by a distance r_{12}, is*

$$V(r) = \frac{z_1 z_2 e^2}{r_{12}} \tag{2.2}$$

The potential arising from the interaction of a charge and a dipole is

$$V(r) = \frac{-z_1 e \mu_2 \cos \theta}{r_{12}^2} \tag{2.3}$$

where μ_2 is the dipole moment of the dipolar molecule 2 and θ is the angle between the direction of the dipole and r_{12}. In practice, Eq. (2.3) can be averaged over all molecular orientations, with the result that $V(r)$ depends on r_{12}^{-4}.

Even if the molecule does not have a permanent dipole moment, an electric field acting on the molecule will induce a moment. The proportionality factor between electric field and dipole moment is the polarizability α, so that

$$\mathbf{\mu} = \alpha \mathbf{E} \tag{2.4}$$

If the molecule is spherically symmetrical, α is a scalar quantity with dimensions of volume. The electric field of a point charge is given by

$$E = -\frac{d}{dr_{12}} \left(\frac{z_1 e}{r_{12}} \right) = \frac{z_1 e}{r_{12}^2}$$

so that the induced dipole is

$$\mu = \frac{\alpha z_1 e}{r_{12}^2}$$

The ionic charge will then interact with this induced dipole, leading to the potential

$$V(r) = \frac{-z_1^2 e^2 \alpha_2}{2r_{12}^4} \tag{2.5}$$

This interaction will be important in a subsequent discussion of ion-molecule reactions. The polarizability can be evaluated experimentally from measurements of the index of refraction. Additivity relations also

---

*This equation and the other electrostatic equations in this chapter apply, strictly speaking, to a vacuum. The potentials should be divided by the dielectric constant of the medium, but this correction is of significance to kinetics only in the liquid phase.

enable one to evaluate this quantity from tables of atomic, group, or bond polarizabilities.

In an analogous way, a dipole will induce a dipole, leading to an attractive interaction potential that is proportional to r_{12}^{-6}. This type of interaction is the physical basis for dispersion forces (sometimes called London forces) between two uncharged molecules without permanent dipole moments. Although a rigorous description of this phenomenon must be quantum mechanical, a qualitative picture is simple. Although the time-averaged electron distribution of molecule 1 must be symmetrical, at a given instant it can be unsymmetrical and hence can lead to an instantaneous dipole moment. This induces a dipole in molecule 2, and the two dipoles interact as described above. Quantum mechanical calculations also predict an interaction potential that is proportional to r_{12}^{-6}.

2.2. Short-Range Repulsive Forces

The considerations of the preceding section are no longer valid for the prediction of interaction potentials when the distance between molecules is small enough for their electron clouds to overlap. In this region, one must turn to the methods of quantum chemistry. It is useful to distinguish between cases where chemical bonding is known to occur (e.g., two polyatomic radicals or two atoms) and where it is not expected to take place (e.g., two stable molecules or two inert gas atoms). In the latter case, the predominant short-range interaction is an extremely strong repulsive potential, which can be empirically represented by forms such as

$$V(r) = Ce^{-mr}$$

or

$$V(r) = Cr^{-n} \tag{2.5}$$

2.3. Empirical Expressions
for Intermolecular Potentials

It is a difficult problem to derive, a priori, an expression for a potential that is accurate for a range of intermolecular distances extending from the repulsive to the attractive region. The problem has been circumvented by adopting various empirical expressions that incorporate some of the known features of interactions between molecules. In each of these models, a compromise must be made between mathematical utility and physical reality. Comparison of certain predictions of these models with the corresponding experimental quantities provides a test of validity and also fills in some of the parameters of the potential energy expressions. The classical comparisons have been with equilibrium properties of gases (virial coefficients) or transport properties (viscosity, diffusion, heat conductivity, and thermal diffusion). More recently, molecular beam scattering experiments have added significantly to our knowledge of intermolecular and interatomic forces (Chapter 3 and Appendix B).

Some of the empirical potentials useful in kinetics are shown in Fig. 2.1 and are described below:

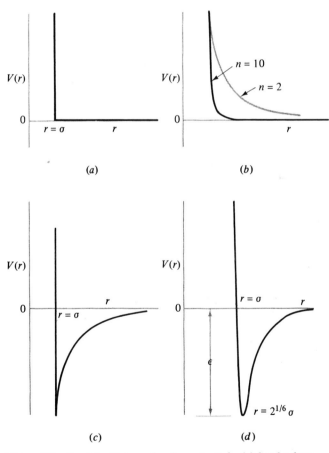

Figure 2.1. *Empirical intermolecular potentials: (a) hard-sphere model; (b) point centers of repulsion; (c) Sutherland model; (d) Lennard-Jones (6-12) potential.*

Hard-Sphere Model. This is represented by

$$V(r) = \infty \qquad r < \sigma$$
$$V(r) = 0 \qquad r > \sigma$$

(2.6)

Mathematically this is the simplest model, and it merely represents rigid, impenetrable spheres of finite volume. It does not account for attractive forces between molecules but does represent short-range repulsive forces.

Point Centers of Attraction or Repulsion. In this case,

$$V(r) = Cr^{-n}$$

(2.7)

If C is positive and n is large (≥ 10), this differs little from the preceding potential. If $n = 2$ and C is positive, we have the potential for two ions of like charge; if $n = 2$ and C is negative, the resultant potential is applicable either to ions of unlike charge or to masses subject to gravitational attraction.

The Sutherland Model. This combines features of each of the preceding models so that

$$V(r) = \infty \qquad r < \sigma$$
$$V(r) = -Cr^{-n} \qquad r > \sigma$$

(2.8)

Both attractive and repulsive forces are included, and as a result there is a minimum in $V(r)$ at the point $r = \sigma$. When $n = 2$, this is a useful model for ion-ion interactions (Chapter 3).

The Lennard-Jones (6-12) Potential. This model is represented by

$$V(r) = 4\epsilon\left[\left(\frac{\sigma}{r}\right)^{12} - \left(\frac{\sigma}{r}\right)^{6}\right]$$

(2.9)

where ϵ is the depth of the potential well and σ is the value of r at which the potential is zero. An attractive potential representing dispersion forces (with the proper r^{-6} dependence) is combined with a steeply rising repulsive potential. By setting dV/dr equal to zero, we find $r = 2^{1/6}\sigma$ for the intermolecular distance at the potential minimum. This model is a fairly realistic and reasonably simple representation of the interaction potential for uncharged molecules, with which a great many calculations and comparisons with experiment have been made.

2.4. Bonding Interactions

When chemical bonding can take place, at distances of a few angstroms, attractive interactions are expected to be greater by orders of magnitude than when chemical bonding is prohibited (Fig. 2.2). The forces between

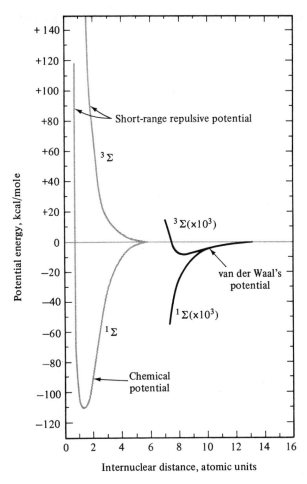

Figure 2.2. *Potential curves for H—H interaction in the bonding ($^1\Sigma$) and non-bonding ($^3\Sigma$) states. The energy is multiplied by a factor of one thousand at large internuclear distances.* [*From H. Pauly and J. P. Toennies,* Advances in Atomic and Molecular Physics **1**, *195 (1965).*]

individual atoms must be taken into account, and we can no longer treat the molecule as a structureless entity. The methods of quantum chemistry must be employed, and formally the situation is the same whether one considers an interaction leading to formation of a stable molecule (e.g., $H + Cl \rightarrow HCl$) or a metathetical process such as $H_2 + Cl \rightarrow HCl + H$. There is not space here for a review of the methods of quantum chemistry, but some of its results will be discussed briefly (2).

2.5. Diatomic Molecules

Potential energies for two-atom systems can be calculated by a number of quantum mechanical techniques. Energies obtained by the best present-day calculations are approaching a precision (~ 1 kcal/mole) that makes them chemically interesting.

A useful empirical expression for a bonding potential is provided by the Morse function:

$$V(r) = D[\exp(-2\beta r) - 2 \exp(-\beta r)] \tag{2.8}$$

where D is the dissociation energy (measured from the bottom of the potential well) and r is the displacement from the equilibrium bond length—that is, $r = R - R_0$. The parameter β is defined in terms of quantities that can be obtained from spectroscopic measurements:

$$\beta = \omega\left(\frac{\mu}{D}\right)^{1/2} \tag{2.9}$$

In this expression, ω is the vibrational frequency and μ is the reduced mass, $m_A m_B/(m_A + m_B)$. The form of this potential (Fig. 2.3) can be

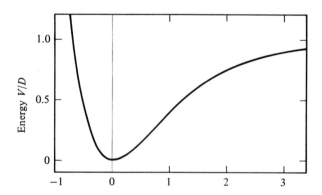

Figure 2.3. *Morse potential for a diatomic molecule. The reduced energy (V/D) is plotted as a function of reduced displacement from the equilibrium interatomic distance ($\beta[R - R_0]$).*

compared with actual potentials derived from observations of vibrational energy levels, and agreement is reasonably good.

2.6. *Polyatomic Molecules*

The quantum mechanical calculation of electronic energies of polyatomic molecules differs from that of diatomic molecules more in quantity than in quality; that is, the computational effort goes up as some power of the number of electrons in the molecule. For this reason, only a few small molecules (usually of high symmetry) have been treated in a rigorous way and with the precision obtained for diatomic molecules.

Even the representation of the energy function, once it has been calculated, presents a problem. The energy depends on *all* of the interatomic distances—that is, for an N-atom molecule,

$$V = V(r_{12}, \cdots, r_{1N}, r_{23}, \cdots, r_{2N}, r_{N-1,N}) \qquad (2.10)$$

and V is a function of $N(N-1)/2$ coordinates. Since physical space only provides us with three coordinates, there is no way of representing such a hypersurface without placing constraints on some of the interatomic distances and thus reducing the dimensionality of the problem.

Consider, for example, a triatomic molecule A—B—C with the ABC angle held constant (e.g., the stable linear configuration of CO_2). The dependence of V on the bond lengths R_{AB} and R_{BC} can be represented as a contour map (Fig. 2.4) on which lines of constant energy (equipotential lines) are plotted with R_{AB} and R_{BC} as axes. (The actual three-dimensional representation corresponding to such a map is commonly called a *potential energy surface*.) The lowest point on this surface indicates the most stable molecular configuration; in this example, point 0 at the bottom of an oddly shaped basin. This basin opens out into two valleys at higher energies: the one at upper left corresponds to the system AB + C, the similar one at lower right to A + BC. The profile of each of these valleys (i.e., the intersection of the energy surface with a plane perpendicular to the page) is the potential energy for the diatomic molecule AB or BC, as in Fig. 2.3. Other important features of the surface are (1) the steep walls at values of R_{AB} or R_{BC} less than the equilibrium bond lengths, indicating strong repulsion; and (2) the plateau at large values of R_{AB} and R_{BC}, which represents dissociation into three separated atoms.

Molecular vibrations can be represented by special paths on such a diagram; the arrow labeled σ corresponds to the symmetric stretching vibration (Fig. 2.5), while that labeled ρ corresponds to the antisym-

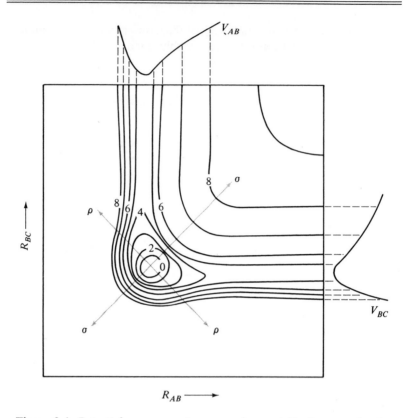

Figure 2.4. *Potential energy contour map for a stable linear molecule, A-B-C. Profiles for dissociation products AB and BC are also indicated. Motion along directions σ and ρ represents the molecular vibrations shown in Fig. 2.5.*

metric stretching mode. Of course, the third allowable vibration (bending away from the linear shape) cannot be shown on this diagram, for only linear configurations are portrayed. The potential energy in the immediate vicinity of the minimum can be expressed in a Taylor series expansion. Again, using the abbreviation $r_{AB} = R_{AB} - (R_{AB})_0$, we have

$$V = V_0 + \left(\frac{\partial V}{\partial R_{AB}}\right)_0 r_{AB} + \left(\frac{\partial V}{\partial R_{BC}}\right)_0 r_{BC} + \left(\frac{1}{2}\right)\left(\frac{\partial^2 V}{\partial R_{AB}^2}\right)_0 r_{AB}^2$$

$$+ \left(\frac{1}{2}\right)\left(\frac{\partial^2 V}{\partial R_{BC}^2}\right)_0 r_{BC}^2 + \left(\frac{\partial^2 V}{\partial R_{AB}\partial R_{BC}}\right)_0 r_{AB} r_{BC} + \text{higher terms} \quad (2.11)$$

The zero subscript in this expression indicates that the quantity is to be evaluated at the equilibrium configuration. The first derivatives are zero

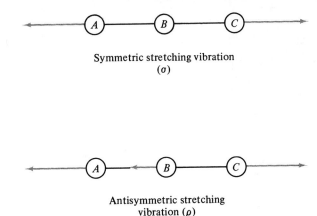

Symmetric stretching vibration
(σ)

Antisymmetric stretching
vibration (ρ)

Figure 2.5. *Stretching vibrations of a linear A-B-C molecule corresponding to lines σ and ρ of Fig. 2.4.*

because the energy is a minimum at the equilibrium point. The second derivatives (or curvatures of the surface) are called force constants, and together with the atomic masses, they allow one to calculate the vibrational frequencies of the motions σ and ρ (3).

Energy contour maps for a few other molecules may be found in the book by Herzberg (4).

2.7. Potential Energy Surfaces
for Reacting Systems

The Arrhenius expression of Chapter 1 provides a hint about the nature of energy surfaces for chemically reacting systems. The term $e^{-E_a/RT}$ suggests that some sort of energy barrier separates reactants from products. The preceding example for triatomic A—B—C will be modified as shown qualitatively in Fig. 2.6. Again, there are valleys corresponding to A + BC and B + AC, as well as a plateau at large values of R_{AB} and R_{BC}. However, the region between valleys is no longer a basin; instead, it is a pass or saddlepoint. Motion of a point along the dashed line (the *reaction path*) represents the configuration of A—B—C throughout the course of the reaction A + BC → AB + C (or its reverse). In the immediate vicinity of the saddlepoint, the reaction path coincides with the *reaction coordinate* (the straight line analogous to ρ

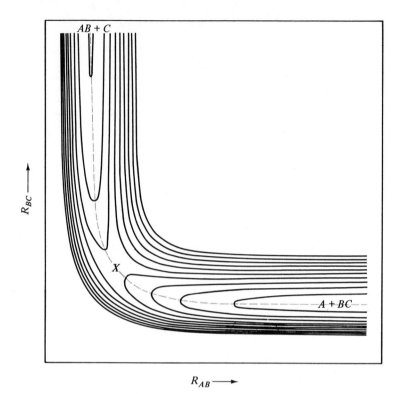

Figure 2.6. *Potential energy contour map for the reacting system $AB + C$ $\rightarrow A + BC$. The reaction path is indicated by the dashed line, which is perpendicular to each contour line it crosses. The saddle point or activated complex is indicated by X.*

in Fig. 2.5); in this same region, the energy can again be expressed in a Taylor series expansion.

The particular configuration at the top of the pass is also called the *activated complex* or the *transition state*, which will be discussed in Chapter 4. The height of this pass above the reactant valley (V_a) is closely related to the activation energy. Although most discussions of potential energy surfaces have been concerned with bimolecular exchange reactions of this type, it is also possible to construct surfaces for unimolecular isomerizations or decompositions (Chapter 5).

The quantum mechanical procedures used to construct one of these surfaces are no different, in principle, from those used to calculate energies of stable molecules. There are some important practical differences, however, which point up the difficulties in applying rigorous *ab initio* methods:

1. In most reactions of the type A + BC → AB + C, A is an atom or a radical; hence ABC has an odd number of electrons, which makes certain quantum mechanical techniques inapplicable.

2. Since ABC is not a stable species, it is not possible to measure properties (energy levels from spectroscopy, for example) which may be compared with calculated values of these same properties. This eliminates an important method of checking the calculations.

3. The barrier height is the difference between two large numbers (the total energy of AB and that of ABC). Small relative errors in the total energies will lead to a large relative error in V_a. The Arrhenius expression shows that a difference of 1.4 kcal/mole in E_a will lead to an order of magnitude change in rate constant at 300 °K.

4. Unlike the case of a stable molecule, where the energy is computed for a small range of interatomic distances around the equilibrium values, a large number of points on the $R_{AB} - R_{BC}$ plane are needed to construct an energy surface useful in kinetics.

Because of these limitations, only a few reaction surfaces have been obtained using rigorous quantum mechanics. That for H_3 (H + para–H_2 → H + ortho–H_2) has been computed by a large number of techniques (5). Calculations have been made for the energies of the ions H_3^+, H_3^{++}, H_4^{++}, and H_4^{3+} over a large range of interatomic distances but with restrictions on the symmetry (6). There also exist surfaces for a few proton-transfer reactions (7):

$$NH_3 + HCl \longrightarrow NH_4Cl$$

$$H^- + H_2 \longrightarrow H_3^- \longrightarrow H_2 + H^-$$

$$H^- + H_2O \longrightarrow H_2 + OH^-$$

$$H^- + NH_3 \longrightarrow H_2 + NH_2^-$$

$$H^- + CH_4 \longrightarrow H_2 + CH_3^-$$

Undoubtedly this type of calculation will be extended to other systems in the very near future. However, in the meantime, chemists have attempted to circumvent the problems of exact quantum mechanics by using semiempirical or completely empirical methods (8) to treat atom-transfer reactions.

A semiempirical approach to the construction of potential energy surfaces was originally developed by Eyring and Polanyi (1931), and more recently modified by Sato (1955). It is based on London's extension of the Heitler-London method for diatomic molecules (9) to three-atom systems. From the combined names of its various contributors, it is

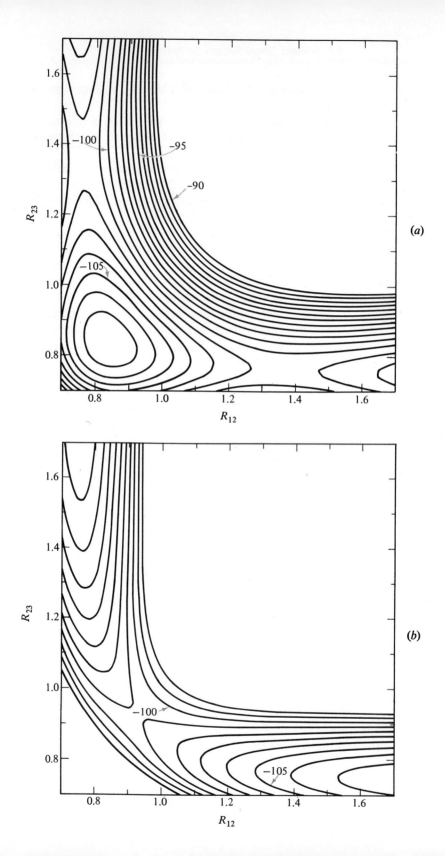

(a)

(b)

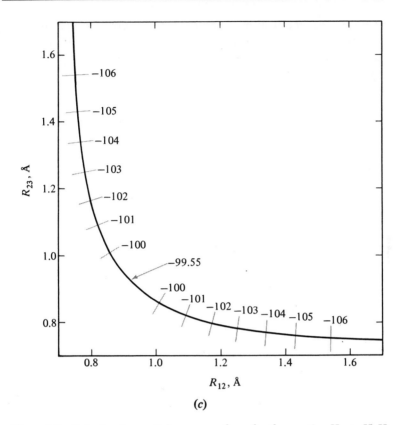

(c)

Figure 2.7. *Calculated potential energy surfaces for the reaction $H_A + H_B H_C$ → $[H_A H_B H_C]$ → $H_A H_B + H_C$:   (a)   from   the   London-Eyring-Polanyi method [Taken from Hirschfelder, Eyring, and Topley,* J. Chem. Phys. **4**, *170 (1936)]; (b) from the Sato modification of this; (c) from the Johnston-Parr BEBO method.*

usually called the LEPS method. (The original 1931 version is the LEP method.) To use this technique, one needs the Morse potential parameters D, R_0, and β for each of the atom pairs (AB, BC, and AC) of the system. In addition, the final energy expression contains an empirical parameter that must be evaluated from some measured property. Usually this parameter is adjusted by a comparison between measured and calculated activation energies. This process uses up one possible prediction, but other features of the potential surface are still of kinetic interest.

A completely empirical method, using no adjustable parameters and requiring no input of kinetic data, has recently been developed by Johnston and Parr (8). Their method has been labeled the BEBO method, because it uses empirical relationships between bond energy and bond order, as well as similar relationships between bond length and bond order. This approach does not lead to a complete surface but only to a reaction path (R_{AB} as a function of R_{BC} throughout the reaction) and the variation of energy along this path. However, the result immediately gives the barrier height V_a and the atomic configuration of the activated complex. Force constants for vibrational motion of the activated complex can also be calculated by the methods outlined by Johnston and Parr. In a comparison of calculated and measured activation energies of a large number of hydrogen-atom transfer reactions, they find agreement within 1 to 3 kcal/mole. In cases where both the LEPS and the BEBO methods have been used, they predict similar properties of the activated complex. Although originally formulated for reactions involving the abstraction of a hydrogen atom by an atom or a radical, the BEBO method has been applied to reactions involving other atom transfers (10).

Potential surfaces obtained by these three methods—LEP, LEPS, and BEBO—applied to the H_3 system are compared in Fig. 2.7. Note that the original LEP surface contains a shallow basin, thus implying the existence of a metastable H_3 species. Neither of the other two surfaces predicts such a basin, nor do any of the more rigorous quantum mechanical calculations.

References

1. Additional details can be found, for example, in J. O. Hirschfelder, C. F. Curtiss, and R. B. Bird, *Molecular Theory of Gases and Liquids*, John Wiley & Sons, Inc., New York, 1954, Chapters 12–14.

2. For background, see (a) K. S. Pitzer, *Quantum Chemistry*, Prentice-Hall, Englewood Cliffs, N. J., 1953 or (b) H. L. Strauss, *Quantum Mechanics: An Introduction*, Prentice-Hall, Englewood Cliffs, N.J., 1968.

3. E. B. Wilson, J. C. Decius, and P. C. Cross, *Molecular Vibrations*, McGraw-Hill, New York, 1955.

4. G. Herzberg, *Electronic Spectra and Electronic Structure of Polyatomic Molecules*, Van Nostrand Reinhold, New York, 1966.

5. The earlier references may be found in General Reference 5 (of Chapter 1), p. 175; the most recent calculation reported is that of I. Shavitt, R. M. Stevens, F. L. Minn, and M. Karplus, *Journal of Chemical Physics* **48**, 2700 (1968).

6. H. Conroy and G. Malli, *J. Chem. Phys.* **50**, 5049 (1969); H. Conroy, *ibid.* **51**, 3979 (1969).

7. For NH_3 + HCl, see E. Clementi, *J. Chem. Phys.* **46**, 3851 (1967). All other examples listed, see C. D. Ritchie and H. F. King, *Journal of the American Chemical Society* **90**, 825, 833, 838 (1968).
8. These are described in some detail in General Reference 5 of Chapter 1.
9. Reference 2(a), p. 128.
10. S. W. Mayer and L. Schieler, *Journal of Physical Chemistry* **72**, 236, 2628 (1968); **73**, 3941 (1969); and earlier references contained therein.

Problems

2.1. At what temperature does the translational energy $(3kT/2)$ equal the electrostatic potential energy of a water molecule (dipole moment 1.8×10^{-18} in centimeter-gram-second units) that is 2 Å from a Fe^{3+} ion?

2.2. The ion-molecule reaction $H_2^+ + H_2 \longrightarrow H_3^+ + H$ is observed to occur in the source of a mass spectrometer. Assume that the kinetic energy of the ion is much larger than that of the molecule and that it results from an accelerating potential of 1 volt. How close to the molecule ($\alpha = 0.79 \times 10^{-24}$ cm^3) must the ion trajectory pass for the kinetic energy and potential energy to be equal?

2.3. Find the vibrational force constant (in centimeter-gram-second units) for a hypothetical Ne_2 molecule if its interatomic potential energy obeys a Lennard-Jones expression with $\epsilon = 49.1 \times 10^{-16}$ erg and $\sigma = 2.77$ Å.

2.4. What is the general expression for the vibrational force constant of a diatomic molecule with potential energy given by the Morse function?

3

Molecular Collisions
and
Nonthermal Reactions

IN THIS CHAPTER we discuss the classical dynamics of collision
processes and show how this subject is related to experimental techniques
like molecular beam scattering and "hot" atom chemistry. The type of
process we consider has two important characteristics.

1. Only two-body forces act during the collision.

2. All the colliding particles in the physical system being considered
have the same velocity.

3.1. Classical Scattering Theory

The classical mechanics needed to describe the scattering of one particle
by another is not unique; it is intimately related to other problems, such
as the description of planetary motion.

To begin with, we consider elastic scattering, in which no energy is lost to internal degrees of freedom of either colliding particle and where the only force is that between the two particles. By a suitable choice of coordinates, the motion of the two-particle system can be transformed into the motion of a hypothetical particle in a central force field.

The coordinate system needed is shown in Fig. 3.1. The positions of

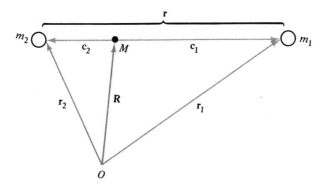

Figure 3.1. *Coordinate system used to describe a two-body collision.*

particles 1 and 2, with masses m_1 and m_2, are defined relative to an arbitrary origin O by vectors r_1 and r_2. The relative position of particle 1 with respect to the second particle is given by another vector

$$\mathbf{r} = \mathbf{r}_1 - \mathbf{r}_2 = \mathbf{c}_1 - \mathbf{c}_2 \tag{3.1}$$

In this expression, the vectors $\mathbf{c}_1$ and $\mathbf{c}_2$ determine the positions of the two particles with respect to the *center of mass*, which is defined by the relation

$$(m_1 + m_2)\mathbf{R} = m_1\mathbf{r}_1 + m_2\mathbf{r}_2 \tag{3.2}$$

$$(m_1 + m_2)\mathbf{R} = m_1(\mathbf{R} + \mathbf{c}_1) + m_2(\mathbf{R} + \mathbf{c}_2) \tag{3.3}$$

Equations (3.1) and (3.3) can be solved simultaneously for $\mathbf{c}_1$ and $\mathbf{c}_2$ in terms of $\mathbf{r}$ and the particle masses, with the result

$$\mathbf{c}_1 = \frac{m_2\mathbf{r}}{(m_1 + m_2)}$$

and

$$\mathbf{c}_2 = -\frac{m_1\mathbf{r}}{(m_1 + m_2)} \tag{3.4}$$

The total kinetic energy of the system is*

$$T = \tfrac{1}{2}(m_1 + m_2)\dot{\mathbf{R}}^2 + \tfrac{1}{2}m_1\dot{\mathbf{c}}_1^2 + \tfrac{1}{2}m_2\dot{\mathbf{c}}_2^2 \tag{3.5}$$

If no external forces act on the particles, the kinetic energy of the center of mass is constant, so that the velocity $\dot{\mathbf{R}}$ is also constant; in particular, it is not changed during the collision. Henceforth we shall neglect this constant contribution to the kinetic energy.

Differentiating (3.4) and substituting into (3.5), we obtain the kinetic energy relative to the moving center of mass

$$T = \tfrac{1}{2}\mu\dot{\mathbf{r}}^2 \tag{3.6}$$

where

$$\mu = \frac{m_1 m_2}{(m_1 + m_2)} \tag{3.7}$$

is the *reduced mass*.

The angular momentum relative to the center of mass is

$$\mathbf{L} = m_1(\mathbf{c}_1 \times \dot{\mathbf{c}}_1) + m_2(\mathbf{c}_2 \times \dot{\mathbf{c}}_2) \tag{3.8}$$

Again we substitute for $\mathbf{c}_1, \mathbf{c}_2, \dot{\mathbf{c}}_1$, and $\dot{\mathbf{c}}_2$ in terms of $\mathbf{r}$ and $\dot{\mathbf{r}}$ to obtain

$$\mathbf{L} = \mu(\mathbf{r} \times \dot{\mathbf{r}}) \tag{3.9}$$

Thus the two-particle problem has been transformed to the motion of a hypothetical particle with mass μ, with its position relative to a fixed scattering center defined by the vector $\mathbf{r}$. This new coordinate system is referred to as the center-of-mass system (since it moves along with the center of mass), but note that the actual particle coordinates with respect to the center of mass are $\mathbf{c}_1$ and $\mathbf{c}_2$.

The angular momentum is another quantity that does not change during the collision. To show this point, we use Newton's equations of motion, again noting that the only forces acting are those of the particles on each other. Hence

$$m_1\ddot{\mathbf{r}}_1 = \mathbf{F}_1 = -\mathbf{F}_2 = -m_2\ddot{\mathbf{r}}_2 \tag{3.10}$$

Therefore

$$\ddot{\mathbf{r}}_1 - \ddot{\mathbf{r}}_2 = \ddot{\mathbf{r}} = \frac{\mathbf{F}_1}{m_1} - \frac{\mathbf{F}_2}{m_2} = \left(\frac{1}{m_1} + \frac{1}{m_2}\right)\mathbf{F}_1 \tag{3.11}$$

and

$$\mu\ddot{\mathbf{r}} = \mathbf{F}_1 \tag{3.12}$$

Recall that $\mathbf{F}_1$ and $\mathbf{F}_2$ act along the line connecting particles 1 and 2—that is, along $\mathbf{r}$. Next, we form the vector product of (3.12) with $\mathbf{r}$ to

---

*We are using the customary notation for a derivative with respect to time: $\dot{x} = dx/dt$.

obtain

$$\mathbf{r} \times (\mu \ddot{\mathbf{r}}) = \mu(\mathbf{r} \times \ddot{\mathbf{r}})$$
$$= \mathbf{r} \times \mathbf{F}_1 = 0 \tag{3.13}$$

The last vector product is zero because $\mathbf{r}$ and $\mathbf{F}_1$ are collinear vectors, and

$$|\mathbf{v}_1 \times \mathbf{v}_2| = v_1 v_2 \sin (v_1 v_2)$$

But

$$\mathbf{r} \times \ddot{\mathbf{r}} = \left(\frac{d}{dt}\right)(\mathbf{r} \times \dot{\mathbf{r}}) - (\dot{\mathbf{r}} \times \dot{\mathbf{r}}) \tag{3.14}$$

Again

$$\dot{\mathbf{r}} \times \dot{\mathbf{r}} = 0$$

so that

$$\mu(\mathbf{r} \times \ddot{\mathbf{r}}) = \left(\frac{d}{dt}\right)\mu(\mathbf{r} \times \dot{\mathbf{r}}) = 0$$

and

$$\mu(\mathbf{r} \times \dot{\mathbf{r}}) = \text{a constant} = \mathbf{L} \tag{3.15}$$

Hence the angular momentum ($\mathbf{L}$) does not change with time. Since the vector product $\mathbf{r} \times \dot{\mathbf{r}}$ (which defines a plane) is constant during the collision or scattering process, the relative motion of the two particles is constrained to a plane perpendicular to $\mathbf{L}$ (Fig. 3.2). Therefore the equations of motion involve only two coordinates, and the particle trajectories can be represented on a single plane (Fig. 3.3). Note, however,

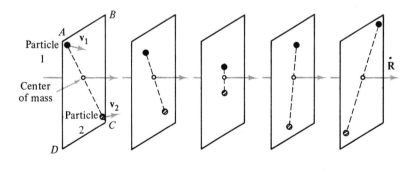

Time =   0 τ 2τ 3τ 4τ

Figure 3.2. *Three-dimensional representation of a two-body collision. This represents a series of snapshots at equally-spaced time intervals. The two particles move on the plane ABCD, which is moving with the center of mass.*

that each individual collision can have a different **L** determined by the absolute directions in space of r_1 and r_2.

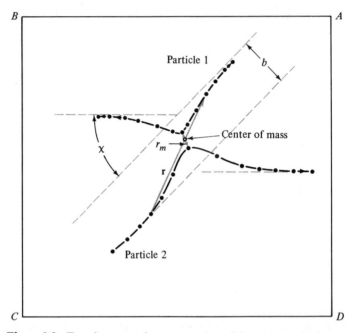

Figure 3.3. *Two-dimensional representation of the same two-body collision as that portrayed in Fig. 3.2. All the ABCD planes have been superimposed and are viewed along the vector which represents motion of the center of mass. The impact parameter b and scattering angle χ are defined as shown.*

3.2. Properties of Trajectories

Let us now consider the details of the scattering process in the new coordinate frame fixed at the scattering center. The vector **r** can be replaced by polar coordinates r and θ (Fig. 3.4). Other quantities useful in the description of the trajectory are

1. The distance of closest approach, r_m.
2. The impact parameter b, which would be the distance of closest approach if no deflection occurred.
3. The scattering angle or angle of deflection, χ.

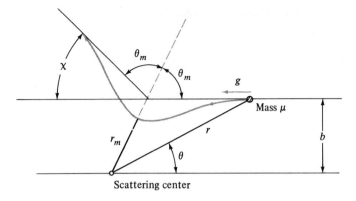

Figure 3.4. *The representation of the same collision as the motion of a hypothetical particle of mass μ with respect to a fixed scattering center. Coordinates used to describe this motion are indicated.*

In this coordinate system, the angular momentum is

$$\mathbf{L} = \mu(\mathbf{r} \times \dot{\mathbf{r}})$$

with magnitude $L = \mu r^2 \dot{\theta}$ (3.16)

Initially r is very large and

$$\theta_0 = \sin \theta_0 = \frac{b}{r_0} \qquad (3.17)$$

$$\dot{\theta}_0 = \frac{-b\dot{r}_0}{r_0^2} = \frac{-bg}{r_0^2} \qquad (3.18)$$

where g is the initial relative velocity (or the velocity of the reduced mass with respect to the fixed scattering center). Since angular momentum is conserved,

$$L = L_0 = -\mu g b \qquad (3.19)$$

Then from (3.19) and (3.16), we have

$$\dot{\theta} = \frac{d\theta}{dt} = \frac{-bg}{r^2} \qquad (3.20)$$

The potential energy of interaction of particles 1 and 2 is $V(r)$, independent of all other coordinates by our initial assumption. Hence the total energy is

$$E = V(r) + \tfrac{1}{2}\mu(\dot{r}^2 + r^2\dot{\theta}^2) = E_0 \qquad (3.21)$$

where E_0 is the initial total energy. In all cases of interest to us, $V(r)$

approaches zero at large distances, so that

$$E_0 = \tfrac{1}{2}\mu g^2 \tag{3.22}$$

Equations (3.20) and (3.21) can be combined to give

$$\dot{r} = \frac{dr}{dt} = \pm g\left[1 - \frac{b^2}{r^2} - \frac{V(r)}{E_0}\right]^{1/2} \tag{3.23}$$

Equations (3.20) and (3.23) are parametric equations that, combined with initial conditions, can be integrated to define values of r and θ as a function of time. Alternatively, they can be combined to give the following differential equation for the trajectory:

$$\frac{d\theta}{dr} = -\frac{b}{r^2}\left[1 - \frac{V(r)}{E_0} - \frac{b^2}{r^2}\right]^{-1/2} \tag{3.24}$$

The negative sign is used, since $d\theta/dr < 0$ for the incoming trajectory. Also, along the incoming trajectory $\dot{r}$ is negative, whereas it is positive along the outgoing trajectory. Therefore at the distance of closest approach (also called the classical turning point of the orbit) $\dot{r}$ is zero, and r_m can be obtained by applying this constraint to Eq. (3.23). The result is

$$r_m = \frac{b}{(1 - V(r_m)/E_0)^{1/2}} \tag{3.25}$$

Note that r_m depends on $V(r_m)$ and not on the potential at any other point along the trajectory. The corresponding value of θ is given by

$$\theta_m = \int_\infty^{r_m}\left(\frac{d\theta}{dr}\right)dr$$
$$= -\int_\infty^{r_m}\frac{b\,dr}{r^2[1 - V(r)/E_0 - b^2/r^2]^{1/2}} \tag{3.26}$$

Because we are considering elastic scattering, the kinetic energy is unchanged by the collision, as is the angular momentum. Hence if it were somehow possible to reverse the path of the particles after the collision took place, it would retrace itself exactly. Therefore the trajectory is symmetrical about the line r_m. From Fig. 3.4 it is evident that the angle of deflection is given by the relationship

$$\chi = \pi - 2\theta_m = \pi - 2b\int_{r_m}^\infty\frac{dr}{r^2[1 - V(r)/E_0 - b^2/r^2]^{1/2}} \tag{3.27}$$

Equations (3.25) and (3.27) show that the trajectory and deflection angle are determined by the initial conditions b and E_0, as well as by the specific form of the interaction potential, $V(r)$. In principle, knowledge of these parameters makes it possible to integrate Eq. (3.27), although doing so may be difficult because of the form of $V(r)$. Some specific cases are discussed by Goldstein (1). Later we shall consider the relation between χ and experimental observables.

3.3. *Approximate Methods for Small Scattering Angles*

The problem of performing the integration of Eq. (3.27) can sometimes be simplified by suitable approximations, particularly when the scattering angle is small ("grazing collisions"). One such method uses the impulse–momentum relationship* (2)

$$\Delta p = \int_{-\infty}^{\infty} F \, dt \tag{3.28}$$

From Fig. 3.5 we see that the scattering angle is

$$\chi \cong \sin \chi = \frac{\Delta p_y}{p} = \frac{1}{p} \int_{-\infty}^{\infty} F_y \, dt \tag{3.29}$$

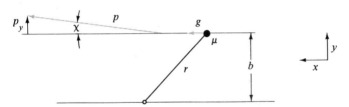

Figure 3.5. *Scattering geometry for collisions producing small deflections.*

and, also from the figure, the component of the force $F(r)$ operating in the y direction is

$$F_y = \frac{y}{r} F(r) \tag{3.30}$$

*This is a useful approach for certain collision processes (e.g., hard-sphere collisions) in which strong forces act for a short time. From Newton's second law,

$$F(r) = \mu \ddot{r}$$

$$\int_{-\infty}^{\infty} F(r) \, dt = \mu \int_{-\infty}^{\infty} \ddot{r} \, dt = \Delta p$$

where Δp is the change in momentum brought about by action of the force $F(r)$. The important point is that although $F(r)$ is very large at the distance of closest approach, it acts during a very short period; hence Δp remains finite.

For the region of interest, $y \cong b$, so that

$$\chi \cong \frac{b}{p} \int_{-\infty}^{\infty} \frac{1}{r} F(r) \, dt$$

and because of the symmetry of the orbit around the point of closest approach

$$\chi \cong \frac{2b}{p} \int_{0}^{\infty} \frac{1}{r} F(r) \, dt \tag{3.31}$$

when the time origin is taken at $r = r_m$. During the incoming segment of the orbit, r is given approximately by

$$r = (b^2 + g^2 t^2)^{1/2} \tag{3.32}$$

from which we may find dt:

$$dt = \frac{r \, dr}{g(r^2 - b^2)} \tag{3.33}$$

Substituting Eq. (3.33) in Eq. (3.31), we have

$$\chi = \frac{2b}{pg} \int_{b}^{\infty} \frac{F(r) \, dr}{(r^2 - b^2)^{1/2}}$$

where the limits of integration arise from the fact that $r \cong b$ when $t = 0$ and $r = \infty$ when $t = \infty$. The initial kinetic energy is given by

$$E_0 = \frac{1}{2} pg$$

Thus

$$\chi = \frac{b}{E_0} \int_{b}^{\infty} \frac{F(r) \, dr}{(r^2 - b^2)^{1/2}} \tag{3.34}$$

is the approximate expression for the scattering angle.

This approximate form is particularly useful for inverse-power potentials of the form

$$V(r) = Cr^{-n}$$

$$F(r) = \frac{-dV(r)}{dr} = Cnr^{-(n+1)} \tag{3.35}$$

In this case,

$$\chi = \frac{bnC}{E_0} \int_{b}^{\infty} \frac{dr}{r^{n+1}(r^2 - b^2)^{1/2}} \tag{3.36}$$

This equation can be integrated by changing variables in the following way:

$$u = \left(\frac{b}{r}\right)^2 \qquad dr = -\frac{1}{2} bu^{-3/2} \, du \tag{3.37}$$

whence

$$\chi = \frac{nC}{2b^n E_0} \int_0^1 u^{(n-1)/2}(1 - u)^{-1/2} \, du \tag{3.38}$$

This definite integral is a standard form expressible in terms of gamma functions:

$$\int_0^1 x^{m-1}(1 - x)^{n-1} \, dx = \frac{\Gamma(m)\Gamma(n)}{\Gamma(m+n)} \tag{3.39}$$

Finally, we have

$$\chi = \frac{C\pi^{1/2}\Gamma(n + 1/2)}{b^n E_0 \Gamma(n/2)} = \frac{CK_n}{b^n E_0} \tag{3.40}$$

Values of K_n for frequently used potentials are $K_1 = 1$, $K_2 = \pi/2$, $K_3 = 2$, $K_6 = 15\pi/16$. Equation (3.40) shows that for large values of n (as in either the repulsive or attractive parts of the Lennard-Jones potential), the scattering angle is very sensitive to the value of the impact parameter.

The famous example of α-particle scattering by an atomic nucleus (Rutherford, 1911) can be treated by this method. The coulomb potential for the interaction between the α particle (charge $2e$) and the nucleus (charge $z_N e$) is

$$V(r) = \frac{2z_N e^2}{r}$$

so that

$$C = 2z_N e^2 \tag{3.41}$$

and $n = 1$. Hence for small scattering angles

$$\chi = \frac{2z_N e^2}{bE_0} \tag{3.42}$$

The more general expression for the scattering angle, which may be found from exact integration of (3.27), is (1)

$$\cot\left(\frac{\chi}{2}\right) = \frac{bE_0}{z_N e^2} \tag{3.43}$$

Note that this same expression applies to the relative motion of two bodies (the sun and a planet, for example) attracted by gravity. In this case, $C = -Gm_1 m_2$, where G is the gravitational constant.

3.4. *The Effective Potential*

We have already been able to simplify the two-body scattering problem by treating it as the motion of a hypothetical particle of mass μ with respect to a scattering center moving at a uniform velocity through space. By using the principle of conservation of angular momentum, it is possible to treat the motion as a hypothetical one-dimensional motion. Once again Eq. (3.21) is

$$E = \tfrac{1}{2}\mu\dot{r}^2 + \tfrac{1}{2}\mu r^2\dot{\theta}^2 + V(r)$$

which may be written as

$$E = \tfrac{1}{2}\mu\dot{r}^2 + V_{\text{eff}} \tag{3.44}$$

The *effective potential* is

$$V_{\text{eff}} = V(r) + \frac{1}{2}\mu r^2\dot{\theta}^2 = V(r) + \frac{L^2}{2\mu r^2} \tag{3.45}$$

Using this expression and the relation $L = \mu bg$, we can rewrite Eq. (3.27) as

$$\chi = \pi - 2b \int_{r_m}^{\infty} \frac{dr}{r^2[1 - V_{\text{eff}}/E_0]^{1/2}} \tag{3.46}$$

By means of Eq. (3.44), one can think of the relative motion as one-dimensional motion in a coordinate r, subject to the condition

$$\dot{r}^2 = 2\frac{(E_0 - V_{\text{eff}})}{\mu} \tag{3.47}$$

Since $\dot{r}^2$ cannot be less than zero, a physically possible trajectory requires that $V_{\text{eff}} \le E_0$. The turning point is reached when $V_{\text{eff}} = E_0$ and $\dot{r}$ is zero.

3.5. Scattering of Molecules that Obey a Lennard-Jones Potential

Some of the foregoing principles can be illustrated for the special case of molecules that interact according to a Lennard-Jones potential. This case is extensively discussed in a number of references (3–5) and is a good representation of physical reality if chemical bonding between the molecules does not interfere.

It is convenient to use dimensionless parameters defined in terms of the Lennard-Jones quantities σ and ϵ as follows:

$$r^* = \frac{r}{\sigma} \qquad b^* = \frac{b}{\sigma}$$

$$V^* = \frac{V}{\epsilon} \qquad E^* = \frac{E_0}{\epsilon} = (g^*)^2 \tag{3.48}$$

$$L^* = \frac{L}{(2\mu\epsilon\sigma^2)^{1/2}} = b^*g^*$$

$$V_{\text{eff}}^* = V^* + \left(\frac{L^*}{r^*}\right)^2 = V^* + E^*\left(\frac{b^*}{r^*}\right)^2$$

In these units, the Lennard-Jones potential is simply

$$V^* = 4[(r^*)^{-12} - (r^*)^{-6}] \tag{3.49}$$

The reduced effective potential is the resultant of two terms, as shown in Fig. 3.6, and there is a family of such curves which depend on the

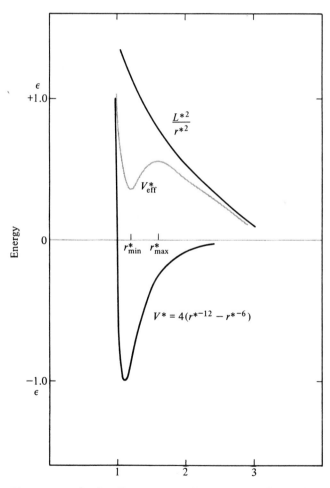

Figure 3.6. V_{eff}^*, *the effective Lennard-Jones potential (in reduced units), and its components* V^* *and* $(L^*/r^*)^2$.

reduced angular momentum L^* (Fig. 3.7). The qualitative shape is also a function of L^*:

1. If $L^* \geq 1.569$, V^*_{eff} is repulsive for all values of r^*.
2. If $L^* < 1.569$, the attractive part of the potential $(-r^{*-6})$ overcomes

the centrifugal repulsion (L^{*2}/r^{*2}) to produce a potential well for some values of r^*. The maximum in V^* at r^*_{max}, a distance larger than that

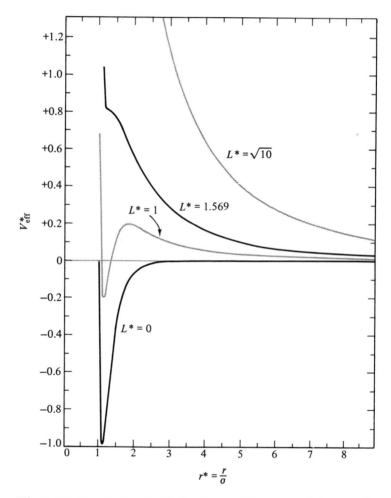

Figure 3.7. *The family of effective Lennard-Jones potential curves for different values of the angular momentum L^*. [From Hirschfelder, Curtiss, and Bird,* Molecular Theory of Gases and Liquids, *Wiley, New York, 1954, p. 554.]*

of the potential well, is called the *centrifugal barrier*. As r^* decreases below the value r^*_{min} at the energy minimum, the term r^{*-12} becomes dominant and repulsion sets in.

The type of collision that takes place can be discussed with the aid of the hypothetical one-dimensional treatment of the preceding section. For a particular value of L^*, which determines the appropriate V^*_{eff} from Fig. 3.7, the trajectory depends on E^* [as indicated in Eq. (3.46)]. The energy is related, in turn, to the impact parameter by

$$E^* = \left(\frac{L^*}{b^*}\right)^2 \tag{3.50}$$

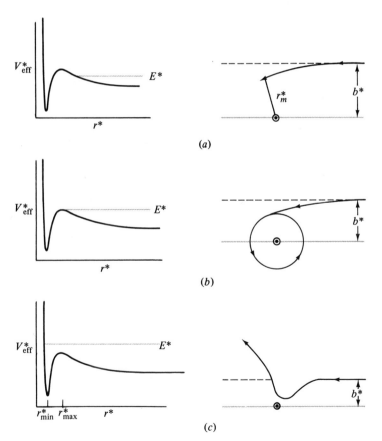

(a)

(b)

(c)

Figure 3.8. *Different types of trajectory for collisions at a fixed L*, but different values of E* (and hence, b*).*

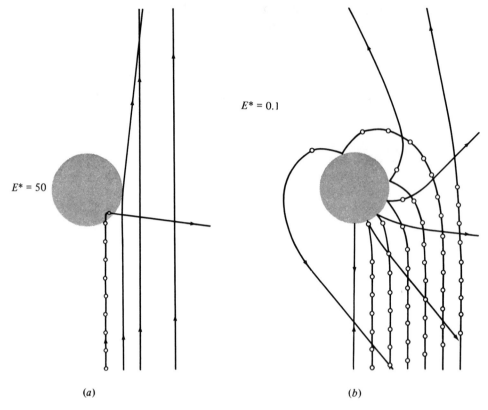

$E^* = 0.1$

$E^* = 50$

(a)   (b)

Figure 3.9. *Actual trajectories for a particle in a central force field corresponding to the Lennard-Jones potential. The shaded disk has a radius equal to the L-J σ. Points along the trajectories indicate positions of the particle at intervals of 0.5 σ/g. (a) $E^* = 50$, (b) $E^* = 0.1$. Thus, the time scale in (a) is smaller by a factor of $(500)^{1/2}$ than that of (b).*

To be specific, let us consider the case where $L^* = 1.2$, and there is a substantial centrifugal barrier. The trajectories fall into three broad classes, represented in Fig. 3.8.

1. If the impact parameter is large and E^* is below the height of the centrifugal barrier, then as r^* decreases so does $|\dot{r}^*|$. The value of $(\dot{r}^*)^2$ is proportional to the distance between the horizontal line corresponding to E^* and the curve for V_{eff}^*. At some value of r^* larger than r_{max}^*, the two lines intersect at the value of r^* corresponding to the turning point. In this example, the deflection is slight.

2. If the value of E^* is exactly equal to V_{eff}^* at r_{max}^*, there is no radial

force at this point, since dV^*_{eff}/dr^* is zero. The particles will orbit indefi-
nitely at this distance.

3. If E^* is larger than V^*_{eff} at r^*_{max}, the velocity $|\dot{r}^*|$ will first decrease
until r^*_{max} is reached, increase again until $r^* = r^*_{min}$, and then decrease
until the turning point is reached at a value of r^* where E^* intersects
the steeply rising repulsive part of V^*_{eff}. The trajectory for this example
of a small impact parameter has a large scattering angle.

Actual trajectories are shown in Fig. 3.9 for a number of impact parameters and for large and small values of E^*. The high-energy trajectories
are similar to those expected for hard-sphere collisions with a collision
diameter of 2σ. The low-energy trajectories are not very different from
the high-energy ones when the impact parameter is small—that is, when
L^* and E^* are both small. At impact parameters near 2σ, the trajectories
become complicated, for the orbiting condition is nearly satisfied.

From calculations of the scattering angle for such trajectories, it is
possible to predict transport properties (viscosity, diffusion, and thermal
conductivity) of a gas.

3.6. *Scattering Cross Sections for Elastic Collisions*

We recall from Eq. (3.27) that the deflection angle χ is determined by
E_0, b, and $V(r)$. Since $V(r)$ is not dependent on the relative orientation
of the two particles, all incoming trajectories that lie on a cylinder of
radius b about the axis $b = 0$ are equivalent (Fig. 3.10). When b is very
large, no deflection takes place; finally, at some value b_{max}, the lower
observable limit (χ_0) of the deflection angle is reached. The effective cross
section for scattering, σ, is the area of the disk of radius b_{max},

$$\sigma = \pi b^2(\chi_0) = \pi b^2_{max} \tag{3.51}$$

Note that σ has units of area.

The definition of total cross section becomes somewhat clearer if one
looks at the specific example of hard-sphere collisions. Here the particles
do not interact unless they are in actual contact. Thus the maximum
impact parameter leading to deflection is the sum of particle radii
(sometimes called the mean collision diameter); that is,

$$b_{max} = r_1 + r_2$$

Therefore, according to Eq. (3.53),

$$\sigma = \pi(r_1 + r_2)^2 \tag{3.52}$$

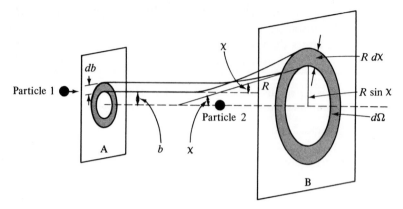

Figure 3.10. *Geometry for the definition of cross section (σ), and differential cross section, $\sigma(\chi)$. The length of line R is arbitrary.*

which is just what one would expect intuitively. Note that there is no dependence of the cross section on the relative velocity of the two particles.

This same result could have been obtained by a more formal route, with the use of Eq. (3.27). Since the distance of closest approach is

$$r_m = r_1 + r_2$$

and

$$V(r) = 0 \qquad \text{for} \qquad r \geq r_1 + r_2 \tag{3.53}$$

the expression for the scattering angle is simply

$$\chi = \pi - 2b \int_{r_1+r_2}^{\infty} \frac{dr}{r(r^2 - b^2)^{1/2}}$$

$$= \pi - 2 \cos^{-1}\left(\frac{b}{r}\right)\Big|_{r=r_1+r_2}^{r=\infty} \tag{3.54}$$

Then

$$\chi = 2 \cos^{-1}\left[\frac{b}{(r_1 + r_2)}\right]$$

or

$$b = (r_1 + r_2) \cos\left(\frac{\chi}{2}\right) \tag{3.55}$$

Again, if χ_0 is zero, then

$$b_{\max} = r_1 + r_2 \tag{3.56}$$

in agreement with our previous result.

The case of an inverse power potential can be treated approximately by means of the small-angle formula previously derived [Eq. (3.40)]. This formula should be a reasonable approximation for large values of E_0, the relative kinetic energy. Rearrangement of Eq. (3.40) gives

$$b^n = \frac{CK_n}{\chi E_0} \tag{3.57}$$

so that

$$b_{max} = \left(\frac{CK_n}{\chi_0 E_0}\right)^{1/n} \tag{3.58}$$

and

$$\sigma(g) = \left(\frac{CK_n}{\chi_0 E_0}\right)^{2/n} = \left(\frac{2CK_n}{\chi_0 \mu g^2}\right)^{2/n} \tag{3.59}$$

Unlike the hard-sphere cross section, this result depends on the initial relative velocity or kinetic energy. However, somewhat to our dismay, we see that as the lower limit of the scattering angle approaches zero, the cross section becomes infinite. This result of the classical calculation follows from the fact that the potential is finite at all values of r, and hence a completely undeflected trajectory is impossible, no matter how large b is. This singularity is removed by quantum-mechanical considerations, which show that σ is finite for potentials that decrease more rapidly than r^{-3} at large values of r.

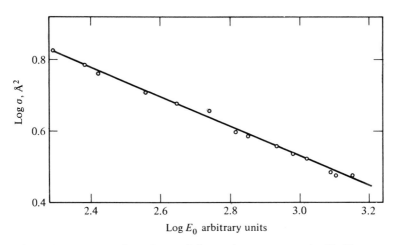

Figure 3.11. *Energy dependence of the total cross section for* He-He *scattering.* [*Data from I. Amdur, J. E. Jordan, and S. O. Colgate,* J. Chem. Phys. **34**, *1525 (1961).*]

In practice, a lower limit is set on χ_0 by the angular resolution of the molecular beam apparatus—that is, the finite size of the detector. Therefore, from Eq. (3.59), $\sigma(g)$ is expected to be proportional to $(E_0)^{-2/n}$ or $g^{-1/n}$. A number of experimental examples of high-energy scattering that are in agreement have been found (6). One such case is He–He scattering in the velocity range of 9 to 24×10^6 cm/sec. From the experimental cross section as a function of energy (Fig. 3.11), Amdur and co-workers (7) derive

$$V(r) = 5.56 \times 10^{-12}\, r^{-5.03} \text{ erg} \qquad (3.60)$$

The range of r values is within the repulsive region of the potential, but note that $V(r)$ is much less sharply dependent on r than would be predicted from a Lennard-Jones expression. However, it is in good agreement with potentials for this system calculated theoretically or determined from the transport properties of helium gas.

3.7. Differential Cross Sections for Elastic Scattering

Figure 3.10 is also useful in defining the differential cross section, which is the cross section for scattering into an element of solid angle at a specific value of χ. A ring-shaped element of solid angle at plane B has the magnitude

$$d\Omega = \frac{\text{area}}{R^2}$$

$$= \frac{2\pi(R \sin \chi)(R\, d\chi)}{R^2} = 2\pi \sin \chi\, d\chi \qquad (3.61)$$

The number of particles passing through the ring $2\pi b\, db$ per unit time is $2\pi Ib\, db$, where I is the flux in particles per unit area per unit time. If we then define the differential cross section $\sigma(\chi)$ according to

$$I\sigma(\chi)\, d\Omega = \text{number of particles scattered into } d\Omega \text{ per unit time then}$$

$$2\pi Ib\, db = I\sigma(\chi)(2\pi \sin \chi\, d\chi) \qquad (3.62)$$

and

$$\sigma(\chi) = \left(\frac{b}{\sin \chi}\right)\left|\frac{db}{d\chi}\right| \qquad (3.63)$$

[The absolute value of $db/d\chi$ is used since $\sigma(\chi)$ is always positive, whereas the sign of $db/d\chi$ depends on whether the potential is attractive or repulsive.] The differential cross section (in units of area/steradian) is related to the total cross section (also known as the integral cross section) by

$$\sigma = \int \sigma(\chi) \, d\Omega = 2\pi \int_{\chi_0}^{\pi} \sigma(\chi) \sin \chi \, d\chi \qquad (3.64)$$

Again, we illustrate this first with the example of hard-sphere collisions. From Eq. (3.55),

$$\left| \frac{db}{d\chi} \right| = \frac{1}{2}(r_1 + r_2) \sin \left(\frac{\chi}{2} \right)$$

and together with Eq. (3.63), this result gives*

$$\sigma(\chi) = \frac{(r_1 + r_2) \cos (\chi/2)}{\sin \chi} \frac{(r_1 + r_2)}{2} \sin \left(\frac{\chi}{2} \right)$$

$$= \frac{1}{4}(r_1 + r_2)^2 \qquad (3.65)$$

There is no angular dependence of the distribution of scattered particles in this case, and it is said to be isotropic. A check on the validity of Eq. (3.65) is provided by (3.64), according to which

$$\sigma = 2\pi \int_0^{\pi} \frac{(r_1 + r_2)^2}{4} \sin \chi \, d\chi$$

$$= \pi(r_1 + r_2)^2 \qquad (3.66)$$

as we found previously.

For an inverse-power potential, the small-angle formulation can again be used. Since the scattering angle is approximately

$$\chi \cong \sin \chi = \frac{K_n C}{E_0 b^n} \qquad (3.40)$$

$$\left| \frac{db}{d\chi} \right| = \left(\frac{d}{d\chi} \right) \left(\frac{CK_n}{E_0 \chi} \right)^{1/n}$$

$$= n^{-1} \left(\frac{CK_n}{E_0} \right)^{1/n} \chi^{-(n+1)/n} \qquad (3.67)$$

The preceding expressions are combined in the form

$$\sigma(\chi) = \left(\frac{CK_n}{E_0 \chi} \right)^{1/n} \left(\frac{1}{\chi} \right) \left[n^{-1} \left(\frac{CK_n}{E_0} \right)^{1/n} \chi^{-(n+1)/n} \right]$$

$$(b) \qquad \left(\frac{1}{\sin \chi} \right) \qquad \left(\frac{db}{d\chi} \right)$$

to give

$$\sigma(\chi) = \frac{1}{n} \left(\frac{CK_n}{E_0} \right)^{2/n} \chi^{-(2n+2)/n} \qquad (3.68)$$

*Note that $\sin 2u = 2 \sin u \cos u$.

Measurements of the cross section $\sigma(\chi)$ as a function of the scattering angle χ should thus lead to a determination of n, the exponent in the expression for the potential energy, $V(r) = Cr^{-n}$. An example is shown in Fig. 3.12 for the scattering of K atoms by Hg atoms at small angles, where $\log \sigma(\chi)$ is plotted against $\log \chi$. The straight-line portion of this plot gives a value of 6 for n, and the interaction is attractive. Measurement of $\sigma(\chi)$ at larger angles enables one to determine the parameters σ and ϵ of the Lennard-Jones potential.

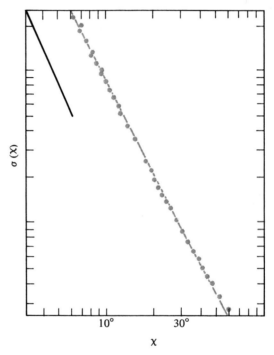

Figure 3.12. *Angular dependence of the differential cross section for* **K-Hg** *scattering at a relative velocity of* 6×10^4 *cm/sec. The line segment at upper left is drawn with a slope of* $-7/3$. [*From F. A. Morse and R. B. Bernstein,* J. Chem. Phys. **37**, *2019 (1962).*]

3.8. Cross Sections for Chemical Reaction

At this point we have developed the concepts necessary to obtain cross sections, and therefore rate constants, for a very limited class of chemical reactions. The limitations are (1) the reaction must be bimolecular; (2) the relative kinetic energy of the reactants must be known; (3) internal degrees of freedom of the reactants do not participate in energy exchange; and (4) reaction must be defined as taking place when the two molecules reach some critical distance R. Even with these severe restrictions, reasonable models for recombination reactions of atoms, radicals, and ions can be devised. Furthermore, this is an essential step toward the derivation of rate constants for systems in thermal equilibrium, which is carried out in the next chapter.

For the example of hard elastic spheres, it only makes sense to define reaction as taking place when the particles are in actual contact. Previously we had obtained the result

$$\sigma = \pi(r_1 + r_2)^2$$

independent of energy. From Appendix B, the rate constant is related to the cross section according to

$$k = \sigma g = \pi(r_1 + r_2)^2 g \qquad (3.69)$$

A more reasonable model for a chemical reaction is one that combines an energy barrier with a hard-sphere model. Again let us assume that reaction occurs when $R = r_1 + r_2$, but that at this point the potential has a value V_R. It is thus a sort of activation energy. At values of r less than $r_1 + r_2$, the potential is infinite, while the form of $V(r)$ at larger distances is immaterial. The largest impact parameter leading to reaction will be one that gives $r_1 + r_2$ as the distance of closest approach; from Eq. (3.25) we have

$$b_{\max}^2 = (r_1 + r_2)^2 \left(1 - \frac{V_R}{E_0}\right)$$

$$\sigma = \pi(r_1 + r_2)^2 \left(1 - \frac{V_R}{E_0}\right) \qquad (3.70)$$

$$k = \pi(r_1 + r_2)^2 g \left(1 - \frac{2V_R}{\mu g^2}\right)$$

which is valid for $E_0 \geq V_R$. In this example, the cross section is zero when E_0 and V_R are equal and increases with energy, approaching the hard-sphere value when $E_0 \gg V_R$.

The cross section for recombination of ions can be treated in a similar way. Previously we indicated that the total cross section derived from a central force potential is infinite, since all trajectories are deflected. However, if we require the two ions to approach to some critical distance R before reaction can take place, the argument of the preceding paragraph can be used. Thus

$$b_{\text{max}}^2 = R^2\left(1 - \frac{V_R}{E_0}\right)$$

where

$$V_R = \frac{z_1 z_2 e^2}{R}$$

Hence

$$\sigma = \pi R^2\left(1 - \frac{z_1 z_2 e^2}{E_0 R}\right) \tag{3.71}$$

$$k = \pi R^2 g\left(1 - \frac{2z_1 z_2 e^2}{\mu g^2 R}\right)$$

The last term (positive for ions of unlike sign) is much larger than unity, so that the attractive potential increases the cross section greatly over the hard-sphere value.

For the more general case of an inverse-power attractive potential,

$$V(r) = -Cr^{-n} \qquad \text{with } n > 2$$

the type of trajectory to be expected has already been discussed for the specific example of a Lennard-Jones potential. It depends on the relative values of E_0 and V_{eff} at the position of the centrifugal barrier. A reasonable assumption is that the maximum value of b leading to a reactive encounter is the value that produces an orbiting trajectory. When this step occurs, dr/dt is zero, and it follows from (3.44) that dV_{eff}/dr is also zero. From Eq. (3.45) we obtain

$$\begin{aligned}\left(\frac{dV_{\text{eff}}}{dr}\right)_{r=R} &= \frac{nC}{R^{n+1}} - \frac{L^2}{\mu R^3}\\ &= \frac{nC}{R^{n+1}} - \frac{(\mu b_{\text{max}} g)^2}{\mu R^3} = 0\end{aligned} \tag{3.72}$$

and from Eq. (3.47)

$$\dot{R}^2 = g^2 - \frac{(g b_{\text{max}})^2}{R^2} + \frac{2C}{\mu R^n} = 0 \tag{3.73}$$

Simultaneous solution of these two equations for R and b_{max}^2 (involving considerable algebraic manipulation) leads to

$$b_{\text{max}}^2 = \left[\frac{C(n-2)}{2E_0}\right]^{2/n} \frac{n}{n-2} \tag{3.74}$$

and

$$R = \left[\frac{C(n-2)}{2E_0}\right]^{1/n} \tag{3.75}$$

As an example, consider the reaction of an ion and a molecule where the interaction is that of a charge and an induced dipole. Then

$$V(r) = \frac{-z_1^2 \alpha_2 e^2}{2r^4}$$

and

$$b_{max}^2 = 2\left(\frac{z_1^2 \alpha_2 e^2}{2E_0}\right)^{1/2} \tag{3.76}$$

Hence

$$\sigma(g) = 2\pi z_1 e \left(\frac{\alpha_2}{2E_0}\right)^{1/2} = 2\pi z_1 e \left(\frac{\alpha_2}{\mu g^2}\right)^{1/2} \tag{3.77}$$

and

$$k = 2\pi z_1 e \left(\frac{\alpha_2}{\mu}\right)^{1/2}$$

This interesting result predicts that the rate constant is independent of the initial relative velocity of the ion and molecule.

3.9. *Reactive Scattering of Particles with More Complex Potentials*

Up to this point we have considered only the particle dynamics of structureless atoms or molecules moving in a potential that depends only on the intermolecular distance. As pointed out in Chapter 2, this approach is a reasonable one when no specific chemical bonding takes place, but it is a very crude approximation for systems in which chemical forces are important. Thus a more realistic approach requires

1. A quantum chemical derivation of a potential energy surface as a function of all interatomic distances (Chapter 2).
2. Solution of the classical equations of motion for each atom involved in the reaction.

Quite a few calculations of this general type (8–14) have been carried out.* We illustrate these with a specific example, the reaction

$$H_A + H_B H_C \longrightarrow H_A H_B + H_C$$

as treated by Karplus, Porter, and Sharma (15, 16). The potential surface they use is obtained from a more sophisticated form of the LEPS method (17), and strongly resembles that of Fig. 2.7(b).

The classical equations of motion have as variables the nine Cartesian

---

*An historical note: The first such calculation was begun by J. O. Hirschfelder in 1936, using a desk calculating machine. The potential energy surface of Fig. 2.7(a) was used. So far as we know, the next such calculation awaited the advent of high-speed digital computers, and was not done until 1958.

coordinates and associated momenta of the three atoms. In principle, by working in the CM system, one could set up the problem in terms of three internal coordinates. Mathematically, it is simpler to work in a system of six coordinates and associated momenta. This process leads to 12 coupled differential equations, 6 of them describing the time dependence of the coordinates, the other 6 that of the momenta. The classical motion of the three atoms is calculated by integrating these equations from initial conditions where R_{AB} is large, to final conditions where one of the atoms is again far away from the hydrogen molecule.

A major complexity of the calculation, which we shall not discuss, concerns the choice of initial values for the six coordinates and momenta. (The initial coordinate values also determine the impact parameter.) In establishing starting values of the momenta, the total angular momentum is quantized. So is the vibrational and rotational energy of the H_2 molecule (quantum numbers v and J, respectively). No quantum restrictions are placed on these properties during the integration procedure, however.

The final state is reached when the atom and molecule are far enough apart so that the interaction potential is small. Collisions between H_A and $H_B H_C$ may be either reactive or nonreactive (Fig. 3.13). The reaction probability or opacity (P_r) is calculated by obtaining a large number of trajectories, such as those illustrated, for an assortment of initial conditions. Then $P_r(g, b, v, J)$ is simply the fraction of these trajectories in which an atom transfer takes place. As Fig. 3.13 indicates, the time period during which the three atoms are close enough to interact is of the order of a molecular vibration period (10^{-14} to 10^{-13} sec).

Finally, the cross section as a function of relative velocity can be obtained from the integral

$$\sigma(g, v, J) = 2\pi \int_{b=0}^{b_{max}} P_r(b, g, v, J) b \, db \qquad (3.78)$$

where b_{max} is the impact parameter at which P_r drops to zero. The form of σ at low kinetic energy is shown in Fig. 3.14(a), and an extension of these calculations to $T + H_2$ and $T + D_2$ collisions (16) yields the larger energy range in Fig. 3.14(b). The decrease in σ at high energies may be interpreted as the result of encounters in which the kinetic energy is so great that it overwhelms the chemical bonding energy. The lowest energy at which σ begins to rise from zero is called the *threshold*.*

---

*We emphasize at this point that there are three similar, frequently confused, but fundamentally different quantities: (1) the threshold energy of a reaction, described above; (2) the barrier height, V_a, of the potential energy surface; and (3) the activation energy, E_a, which is defined empirically in terms of the Arrhenius expression but which can also be defined in a more fundamental way (Section 5.5).

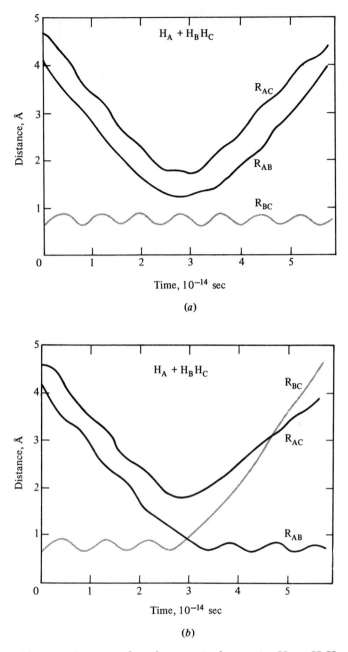

Figure 3.13. *Internuclear distances in the reaction* $H_A + H_B H_C$ *as a function of time:* (*a*) *a non-reactive collision; the small fluctuations in* R_{BC} *are due to molecular vibration;* (*b*) *a reactive collision; crossing of* R_{AC} *and* R_{BC} *at* 4×10^{-14} *secs indicates a rotation of* $H_A H_B$. [*From Karplus, Porter, and Sharma*, J. Chem. Phys. **43**, *3259 (1965)*.]

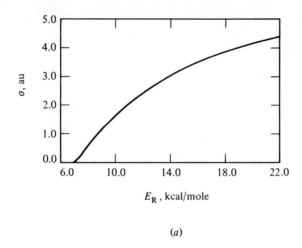

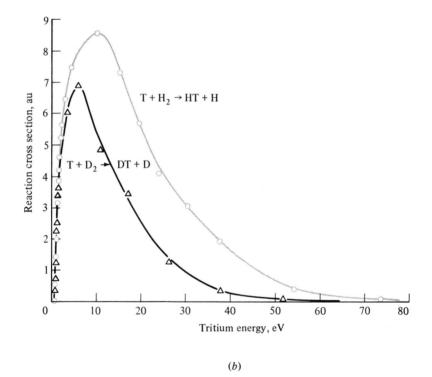

Figure 3.14. (a) *energy dependence of the cross section* $\sigma(g, v = 0, J = 0)$ *for* H + H$_2$ [*from the same source as Fig. 3.13*]; (b) *analogous cross sections for* T + H$_2 \rightarrow$ TH + H *and* T + D$_2 \rightarrow$ TD + D. [*from Karplus, Porter, and Sharma*, J. Chem. Phys. **45**, 3871 (1966)]. *In both of these,* σ *is in atomic units* (1 a. u. = 0.2798 Å^2).

3.10. Experimental Determination of Cross Sections for Nonthermal Reactions

HOT ATOM TECHNIQUES

We discuss first three sets of experiments designed to provide an experimental comparison with the theoretical results just described.

The first is a measurement of the threshold energy for the reaction $D + H_2 \rightarrow DH + H$ (18). In essence, the experiment involves the production of D atoms with a kinetic energy much greater than the average thermal energy of their surroundings. This process is accomplished by the photolysis of a hydrogen halide:

$$DI + h\nu \longrightarrow D^* + I \qquad (D^* = \text{``hot'' D atom}) \qquad (3.79)$$

If the quantum of light has energy greater than the bond-dissociation energy, $D(DI)$, the D and I atoms will be produced with a total relative kinetic energy equal to

$$E_0 = D(DI) - h\nu$$
$$= \tfrac{1}{2}\mu g^2 = \tfrac{1}{2}m_D \dot{c}_D^2 + \tfrac{1}{2}m_I \dot{c}_I^2 \qquad (3.80)$$

From Eq. (3.4), this energy must be divided among the two atoms in such a way that

$$\dot{c}_D = \frac{m_I g}{m_I + m_D} \qquad \dot{c}_I = \frac{m_D g}{m_I + m_D}$$

and

$$E_D = \frac{1}{2}m_D \dot{c}_D^2 = \frac{m_I E_0}{m_D + m_I} \qquad (3.81)$$

Thus the lighter atom receives almost all of the kinetic energy released when the DI molecule dissociates, and it is called a "hot" atom. In the presence of added hydrogen, there are two main paths of reaction open to the D* atom:

Hydrogen atom abstraction, $D^* + H_2 \xrightarrow{k_1} DH + H$ (3.82)

or

Loss of kinetic energy by collision, $D^* + H_2 \xrightarrow{k_2} D + H_2$ (3.83)

Followed by $D + DI \xrightarrow{k_3} D_2 + I$ (3.84)

An argument similar to that used in deriving Eq. (3.81) indicates that the energy available in the CM coordinate system for the $D + H_2$ encounter is

$$E = \frac{m_{H_2} E_D}{m_{H_2} + m_D} \simeq \frac{E_0}{2} \tag{3.85}$$

It is known that the activation energy for reaction (3.84) is very low compared with that of (3.82). Hence most of the thermal D atoms will disappear by reaction with DI, whereas most of the hot atoms will disappear in steps (3.82) and (3.83). Thus a determination of the (DH)/(D$_2$) ratio should yield the rate constant ratio k_1/k_2. (The complete kinetic analysis is somewhat more complicated than this.) By illuminating the DI-H$_2$ mixture with light of different wavelengths, it is possible to obtain the ratio k_1/k_2 as a function of energy. If the energy of D* were sharply defined, and if k_2 were known as a function of energy, it would be possible to obtain the energy dependence of k_1 (and the analogous cross section). Although this goal has not yet been achieved, the threshold energy of the reaction has been obtained by finding the energy at which reaction (3.82) has become negligible. This procedure is shown in Fig. 3.15. The value of 6.0 kcal/mole may be compared with the theoretical threshold of 5.69 kcal/mole.

Related experiments have been done by Chou and Rowland (19), who photolyzed TBr (T is a weakly radioactive isotope of hydrogen) in the

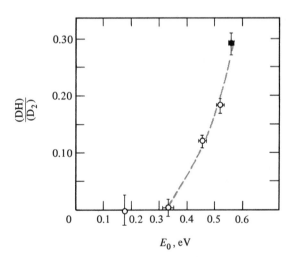

Figure 3.15. *The ratio* (DH)/(D$_2$) *produced by photolysis of deuterium halides in the presence of* H$_2$. *Open circles are from photolysis of* HI *at 3660, 3130, and 3030 Å; full circle is from photolysis of DBr at 2537 Å.* [*From Kuppermann and White, J.* Chem. Phys. **44**, *4532 (1966).*]

presence of HD, or of H_2-D_2 mixtures. The tritium atoms were produced with 3 eV of kinetic energy, and the important reactions are

$$T^* + H_2 \xrightarrow{k_{H_2}} TH + H$$

and

$$T^* + D_2 \xrightarrow{k_{D_2}} TD + D$$

or

$$T^* + HD \xrightarrow{k_{HD}} TH + D$$

and

$$T^* + DH \xrightarrow{k_{DH}} TD + H$$

The hydrogen was then analyzed to determine the (TH)/(DH) ratio, and hence k_{H_2}/k_{D_2} or k_{HD}/k_{DH}. It was found that

$$\frac{k_{H_2}}{k_{D_2}} \simeq 1.0$$

and

$$\frac{k_{HD}}{k_{DH}} \simeq 0.7$$

The first ratio agrees with the predictions shown in Fig. 3.14(b), whereas the second ratio has not yet been calculated theoretically. Note that this experiment circumvents the need to know the nonreactive cross section for thermalization of hot T atoms.

In addition to photodissociation, nuclear reactions can be used to produce hot atoms, for example

$$^3He + neutron \longrightarrow T^* + proton + 1.9 \times 10^5 \text{ eV}$$

This technique has been used by Wolfgang and co-workers (20) to study the reaction just described. Since the hot tritium atom is produced with energies far above the high-energy threshold of Fig. 3.14(b), a number of collisions in which kinetic energy is lost must take place before the atom becomes reactive. Then there is a competition between reaction and energy loss until the kinetic energy of the hot atom has fallen below the lower threshold. Therefore the quantity of TH produced is related to the integral

$$I_R = \int_{E_1}^{E_2} \sigma(E) \, dE$$

where E_1 and E_2 are the upper and lower values of E at which σ is zero. Absolute values of this integral depend on the model used to predict the rate of energy loss. However, as in the photochemical experiments, a ratio $I_R(H_2)/I_R(D_2)$ can be obtained more directly. The value of 1.15 is in reasonable agreement with the relative areas (1.37) of the corresponding curves in Fig. 3.14(b).

Hot atoms or radicals (produced by photolysis or by nuclear reactions) provide a method of exploring chemical reactions that are inaccessible in systems at thermal equilibrium, even at very high temperatures. Several extensive reviews on the subject may be found (21).

CROSS SECTIONS FROM MOLECULAR BEAM EXPERIMENTS

The experimental principles that are the basis for this method are outlined in Appendix B, which should be read at this point.

The most direct approach to the determination of a cross section for chemical reaction between AB and C would be to fill a scattering chamber with AB and to measure the attenuation of a beam of C atoms, using the total cross-section technique of Section B.1. This procedure is not feasible, however, because most of the scattering is nonreactive and it will mask the reactive scattering.

A succesful method involves measurement of the angular distribution of the reaction product, in this hypothetical case, AC or B. The differential reactive cross section thus obtained can be converted into a total cross section (and hence a rate constant) with Eq. (3.64). Until the present, most of these studies (22,23) have been limited to reactions of the general type

$$M + RX \longrightarrow MX + R$$
$$M + X_2 \longrightarrow MX + X$$
$$M + M'X \longrightarrow MX + M'$$

where M is an alkali metal, R an alkyl radical or a hydrogen atom, and X a halogen. This situation, resulting from the use of surface-ionization detectors that are sensitive to M or MX, is expected to change as more universal detectors (such as mass spectrometers) come into use. For example, preliminary results for the reactions $H + D_2$ and $D + H_2$ have been reported (24, 25).

Somewhat surprisingly, it is possible to determine *reactive* cross sections from an analysis of the differential cross section for elastic scattering (26). The principle involved can be seen from a comparison of the intensity at large angles of a scattered K atom beam, when the scattering target is either Kr or the isoelectronic compound HBr (Fig. 3.16). At large angles, the differential cross section $\sigma(\chi)$ becomes constant for the nonreactive case, whereas it continues to decrease with increasing angle when reaction with HBr occurs. It is reasonable to assume that the

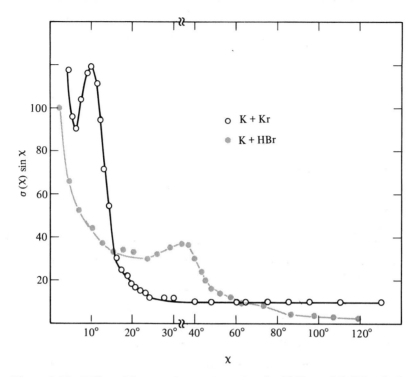

Figure 3.16. *Differential scattering cross section for* **K-Kr** *and* **K-HBr**, *both at* $E_0 = 1.5\ kcal/mole$. *The ordinate scale is in arbitrary units which differ for the two cases.* [*From D. Beck,* J. Chem. Phys. **37**, *2884 (1962).*]

decrease in this case (note that a decrease in elastic cross section means that *fewer* K atoms reach the detector) is due to the additional removal of K atoms by chemical reaction. In fact, from the angular dependence of total scattering at smaller angles (and hence large impact parameters) where chemical reaction does not contribute, it is possible to predict the elastic contribution to the scattering at large angles. The difference between this contribution, $\sigma_{el}(g, \chi)$, and the observed cross section, $\sigma_{obs}(g, \chi)$, can be used to calculate the total reactive cross section $\sigma(g)$ according to

$$\sigma(g) = 2\pi \int [\sigma_{el}(g, \chi) - \sigma_{obs}(g, \chi)] \sin \chi\ d\chi$$

The reactive cross section for $K + HBr \rightarrow KBr + H$ obtained in this way is in good agreement with that determined more directly from measurements of the KBr distribution (27).

3.11. Energy Exchange in Inelastic Collisions

Although we have briefly discussed the measurement of cross sections for chemical reactions, we have not exploited the feature that makes such collisions inelastic—namely, the interconversion of kinetic and internal energy. The latter category includes electronic, vibrational, and rotational energy.

In Appendix B.3 energy conservation is used to obtain the expression

$$\mu' g'^2 - \mu g^2 = 2\,\Delta E = 2(\Delta D - \Delta I) \tag{3.86}$$

where ΔD and ΔI are the changes in bond-dissociation energy and internal energy occurring in the inelastic collision. Experimentally, ΔI can be determined by measuring g and g', provided that ΔD is known from thermochemical data.

MOLECULAR BEAM EXPERIMENTS

Appendix B.3 describes briefly the experimental arrangement for such a study. It is pointed out that angular distribution measurements have often been made, but that determinations of product velocity are also needed for a complete kinematic analysis of the collision. A class of experiments in which both properties have been determined is represented by recent work on ion-molecule reactions (28). As an example, we consider the reaction

$$N_2^+ + D_2 \longrightarrow N_2D^+ + D$$

The primary ions (N_2^+) are produced by electron bombardment of nitrogen, and the beam of N_2^+ is selected by a mass spectrometer. (Fig. 3.17). The beam then enters a cylindrical scattering chamber, about the axis of which an ion detector pivots (in the plane of the drawing). The detector includes an electrostatic "velocity" analyzer, which transmits ions of a given kinetic energy regardless of their mass, and a quadrupole mass filter, which selects the ion of interest. This arrangement enables one to measure the distribution of product ions (N_2D^+) with respect to velocity and scattering angle (both measured in the laboratory coordinate system). The primary ion kinetic energies are much higher than the mean thermal energy of the target molecules; therefore the center of mass velocity vector is almost collinear with the vector $v_{N_2^+}$ and has the magnitude $[m_{N_2^+}/(m_{N_2^+} + m_{D_2})]^{1/2} v_{N_2^+}$ (Fig. 3.18). The velocity (in the CM system) of the product ion can be found from Eqs. (B.10) and

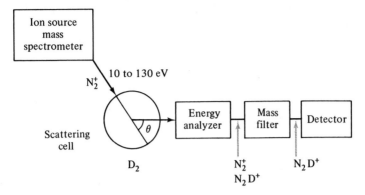

Figure 3.17. *Block diagram of the apparatus used to study the dynamics of ion-molecule reactions. The composition of the ion current at various points is indicated.* [*Mahan,* Accts. Chem. Res. *1, 217 (1968).*]

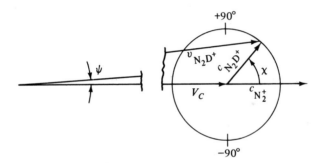

Figure 3.18. *Velocity vector diagram for the reaction* N_2^+ $+ D_2 \longrightarrow N_2D^+ + D$. *The scale of the right-hand side is exaggerated. The quantities measured (in the LAB system) are* $v_{N_2D^+}$ *and* $\psi_{N_2D^+}$.

(B.11) of Appendix B. The result is

$$c_{N_2D^+} = c_{N_2^+}\left(\frac{m_{N_2} \cdot m_D}{m_{N_2D} \cdot m_{D_2}}\right)^{1/2}\left(1 + \frac{\Delta D - \Delta I}{E_0}\right)^{1/2} \tag{3.87}$$

and this velocity can have any direction in space. Thus, in the planar configuration of the experiment, the tip of the corresponding vector describes a circle centered at the CM origin and with a radius given by Eq. (3.87).

Measurement of ψ (the scattering angle in LAB coordinates) and $v_{N_2D^+}$ (the velocity in the LAB system) determines $c_{N_2D^+}$ experimentally.

With a given initial ion velocity, the maximum possible value of $c_{\mathrm{N_2D^+}}$ will be found for the case when ΔI is small—that is, when little of the kinetic energy is transformed to internal energy. Conversely, the lower limit is provided by the value of ΔI that corresponds to dissociation of the product ion. With these requirements and the value of $\Delta D = 1$ eV, the shaded areas in velocity space of Fig. 3.19 are prohibited. The observed ion intensities can be plotted as a contour map in polar coordinates $\chi_{\mathrm{N_2D^+}}$ and $c_{\mathrm{N_2D^+}}$.

Most of the product ions are formed with a velocity near the lower limit, and hence with almost enough internal energy to dissociate. (The spillover of intensity contours into the "forbidden" area of Fig. 3.19 is the result of the finite resolution of the detector and the initial thermal velocity of D_2 molecules.) Additional information is provided by the

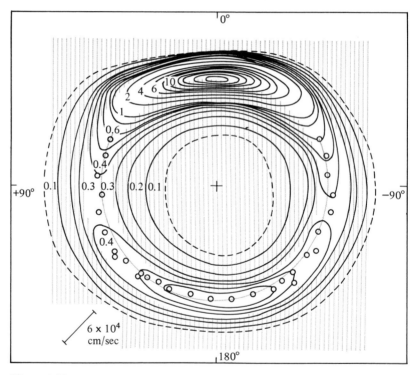

Figure 3.19. *Contour map of* N_2D^+ *ion intensity plotted with polar coordinates* $c_{N_2D^+}$ *and* χ. *In this experiment,* $E_0 = 8.1\,eV$. *Shaded areas represent velocities prohibited by Eq. (3.87) and product stability. Open circles are intensity maxima obtained in energy scans at fixed angle or angle scans at fixed energy. [From B. H. Mahan,* Accounts Chem. Research **1**, *217 (1968).]*

angular distribution of product ions. Suppose that a complex ion $N_2D_2^+$ were formed, stable enough to rotate at least once before breaking up into products. Then the products would "forget" the direction of the original relative velocity and would be equally likely to be scattered in a direction parallel to this velocity or antiparallel to it. Thus the product distribution would be symmetrical about the deflection angle $\chi = 90°$. In fact, a very asymmetrical distribution is found, and the "forward" scattering corresponds to grazing collisions with large impact parameters; this, in turn, leads to a large value of σ, according to Eq. (3.53). The much smaller intensity found near $\chi = 180°$ is the result of "rebound" scattering associated with nearly head-on collisions.

A simplified model that has been used to describe this type of ion-molecule reaction is the "spectator-stripping" model. Let us write a more general hydrogen-atom transfer reaction as

$$A^+ + HB \longrightarrow AH^+ + B$$

The important assumption of this model is that the velocity of atom B (in the CM system) is the same before and after the collision—that is,

$$\dot{c}_B = \dot{c}_{HB}$$

Since this requires that both the magnitude and direction of these vectors be identical, $\dot{c}_{AH}$ must be collinear with $\dot{c}_A$, so that the scattering angle in the CM system is zero. The magnitude of this vector is given by the requirements of conservation of momentum:

$$m_A\dot{c}_A = -(m_B + m_H)\dot{c}_{HB} \quad \text{(initially)}$$

and

$$m_B\dot{c}_B = -(m_A + m_H)\dot{c}_{AH} \quad \text{(finally)}$$

Hence

$$\dot{c}_{AH} = \frac{m_A m_B}{(m_A + m_H)(m_B + m_H)}\dot{c}_A = \frac{m_A m_B}{(m_A + m_B + m_H)(m_A + m_H)}g \tag{3.88}$$

Equation (3.88) must be consistent with Eq. (3.87), and this can be achieved only if all of the bond strength difference, ΔD, plus some translational energy goes into internal energy of the products. The net change in translational energy in the center of mass is given by

$$\Delta E = -\frac{(m_A + m_B + m_H)m_H}{(m_A + m_H)(m_B + m_H)}E_0 \tag{3.89}$$

The idealized stripping model is consistent with observations with H_2 and D_2 as target molecules when the initial relative energies are between 3 and 6 eV. With greater initial energies [as in Fig. (3.19)], the most probable product velocity is higher than that predicted by the preceding expressions. Part of this discrepancy results from the neglect (in this oversimplified model) of the initial velocity of target molecules. At lower energies (~ 1 eV), the model must be modified by the inclusion of forces between the ions and induced dipoles in the molecules, which also lead to greater product velocities.

A contrasting behavior is shown by the similar reaction

$$O_2^+ + D_2 \longrightarrow O_2D^+ + D$$

At low energies, the angular distribution of the product ion is almost isotropic in the CM system (29), thereby indicating formation of a complex that exists for at least a few rotational periods. As the kinetic energy of the incident ion is increased, the complex becomes unstable, and the forward peaking characteristic of the spectator-stripping model is again observed.

CHEMILUMINESCENCE OF PRODUCT MOLECULES

A completely different approach to the determination of energy partition among translational and internal degrees of freedom is provided by the examination of emission spectra of products.* Electronic, vibrational-rotational, and stimulated emission (laser) spectra have been observed. For example, in the reaction

Table 3.1 *Fraction of Available Energy Appearing as Vibrational Excitation of* HCl

v^a	Fraction of total energy, f_v	$P(f_v)^b$
1	0.17	0.25 (I)
		0.31 (II)
2	0.36	0.45 (I)
		(1.00) (II)
3	0.53	(1.00) (I)
		0.85 (II)
4	0.69	0.20 (I)
		0.10 (II)
5	0.83	0.019 (I)
6	0.97	0.0021 (I)

aVibrational quantum number.
bProbability that the fraction f_v goes into vibration in the HCl molecule, normalized to unity for the most probable level. I and II refer to different experimental procedures.

---

*Strictly speaking, this topic belongs in the next chapter, since the reactants are in thermal equilibrium. However, the products are not at the time their spectra are obtained.

$$H + Cl_2 \longrightarrow HCl + Cl + 45 \text{ kcal/mole}$$

infrared emission from as high as the sixth vibrational level of HCl is found (30). For this example, Table 3.1 displays the probability that a given fraction of the available energy will go into vibrational excitation of the product molecule.

COMPARISON WITH POTENTIAL ENERGY SURFACES

So far, results of the type described in this section have not been used to infer fine details of potential energy surfaces for reacting systems. This situation contrasts with elastic scattering, where, for example, the parameters of the Lennard-Jones potential can be determined from experimental results. In the case of reactive scattering, only general features of the relevant potential energy surfaces have been derived. It has been possible, however, to codify several general classes of chemical reaction according to the type of surface (General References 2–4).

References

1. H. Goldstein, *Classical Mechanics*, Addison-Wesley, Reading, Mass. 1959, pp. 71–76.
2. L. D. Landau and E. M. Lifschitz, *Mechanics*, Pergamon Press, Oxford, 1960, p. 55.
3. J. O. Hirschfelder, C. F. Curtiss, and R. B. Bird, *Molecular Theory of Gases and Liquids*, John Wiley & Sons, New York, 1954, pp. 552 ff; J. O. Hirschfelder, R. B. Bird, and E. L. Spotz, *Journal of Chemical Physics*, **16**, 968 (1948).
4. W. Kauzmann, *Kinetic Theory of Gases*, W. A. Benjamin, New York, 1966, pp. 225–230.
5. K. W. Ford and J. A. Wheeler, *Annals of Physics* (*N.Y.*) **7**, 287 (1959).
6. I. Amdur and J. E. Jordan, *Advances in Chemical Physics* **10**, 29 (1966).
7. I. Amdur, J. E. Jordan, and S. O. Colgate, *J. Chem. Phys.* **34**, 1525 (1961).
8. F. T. Wall, L. A. Hiller, Jr., and J. Mazur, *J. Chem. Phys.* **29**, 255 (1958), $H + H_2$ using the LEP potential.
9. D. L. Bunker, *J. Chem. Phys.* **37**, 393 (1962); unimolecular decompositions.
10. N. C. Blais and D. L. Bunker, *J. Chem. Phys.* **39**, 315 (1963); general $A + BC$ reactions.
11. N. C. Blais and D. L. Bunker, *J. Chem. Phys.* **37**, 2713 (1962); $M + CH_3I$.
12. M. Karplus and L. M. Raff, *J. Chem. Phys.* **41**, 1267 (1964); **44**, 1212 (1966); $K + CH_3I$.
13. J. C. Polanyi and S. D. Rosner, *J. Chem. Phys.* **38**, 1028 (1963); $H + Cl_2$.
14. N. C. Blais, *J. Chem. Phys.* **49**, 9 (1968); $M + Cl_2$.

15. M. Karplus, R. N. Porter, and R. D. Sharma, *J. Chem. Phys.* **43**, 3259 (1965); H + H$_2$.
16. M. Karplus, R. N. Porter, and R. D. Sharma, *J. Chem Phys.* **45**, 3871 (1966); T + H$_2$, T + D$_2$.
17. M. Karplus and L. N. Porter, *J. Chem Phys.* **40**, 1105 (1964).
18. A. Kuppermann and J. M. White *J. Chem. Phys.* **44**, 4352 (1966); Proceedings Fifth Nobel Symposium, 1967.
19. C. C. Chou and F. S. Rowland, *J. Chem. Phys.* **46**, 812 (1967).
20. D. Seewald, M. Gersh, and R. Wolfgang, *J. Chem. Phys.* **45**, 3870 (1966).
21. For example, R. Wolfgang, *Annual Reviews of Physical Chemistry*, **16**, 15 (1965); *Progress in Reaction Kinetics* **3**, 99 (1965); *Scientific American* **214**, 82 (1966).
22. D. R. Herschbach, *Advances in Chemical Physics* **10**, 319 (1966).
23. R. B. Bernstein and J. T. Muckerman, *Advances in Chemical Physics* **12**, 389 (1967). Contains an extensive inventory of low-energy molecular-beam scattering results.
24. W. L. Fite and R. T. Brackmann, *J. Chem. Phys.* **42**, 4057 (1965); J. Geddes, H. F. Krause, and W. L. Fite, *ibid.* **52**, 3296 (1970).
25. S. Datz and E. H. Taylor, *J. Chem. Phys.* **39**, 1896 (1963).
26. D. Beck, *J. Chem. Phys.* **37**, 2884 (1962).
27. D. Beck, E. F. Greene, and J. Ross, *J. Chem. Phys.* **37**, 2895 (1962).
28. B. H. Mahan, *Accounts of Chemical Research* **1**, 217 (1968); R. Wolfgang, *ibid.* **2**, 248 (1969).
29. E. A. Gislason, B. H. Mahan, C.-W. Tsao, and A. S. Werner, *J. Chem. Phys.* **50**, 5418 (1969); A. Henglein, *ibid.* **53**, 458 (1970).
30. K. G. Anlauf, P. J. Kuntz, D. H. Maylotte, P. D. Pacey, and J. C. Polanyi, *Discussions of the Faraday Society*, **44**, 183 (1967). Earlier references are given here.

General References

1. R. D. Present, *Kinetic Theory of Gases*, McGraw-Hill, New York, 1958, Chapter 8.
2. H. Pauly and J. P. Toennies, "The Study of Intermolecular Potentials with Molecular Beams at Thermal Energies," in *Advances in Atomic and Molecular Physics*, Bates and Estermann (Eds.), Academic Press, New York, 1965.
3. H. Pauly and J. P. Toennies, "Beam Experiments at Thermal Energies," in *Methods of Experimental Physics*, Bederson and Fite (Eds.), Academic Press, New York, 1968, vol. 7A.
4. J. Ross (Ed.), "Molecular Beams," in *Advances in Chemical Physics*, Interscience Publishers, New York, 1966, vol. 10.
5. E. F. Greene and A. Kuppermann, "Chemical Reaction Cross Sections and Rate Constants," *Journal of Chemical Education* **45**, 361 (1968).
6. R. B. Bernstein, "Molecular Beam Scattering at Thermal Energies," *Science* **144**, 141 (1964).

7. E. F. Greene and J. Ross, "Molecular Beams and Chemical Reactions," *Science* **159**, 587 (1968).

Problems

3.1. A spaceship approaches the moon (mass 7.35×10^{25} g) with an initial relative velocity of 2 km/sec and an impact parameter of 1500 km. The gravitational potential is $V(r) = -Gm_1m_2/r$, with $G = 6.66 \times 10^{-8}$ erg cm g^{-2}.
(*a*) What is the angle of deflection of the spaceship trajectory?
(*b*) The radius of the moon is 1738 km. Find the impact parameter for the spaceship with the velocity given above which would lead to a grazing collision with the moon.

3.2. The total cross section for scattering of K atoms by Ar atoms has been determined by the molecular beam method [E. W. Rothe and R. B. Bernstein, *J. Chem. Phys.* **31**, 1619 (1959)]. In a typical experiment, the following conditions prevail:

K beam current, 3×10^{-10} amp
Effective length of scattering chamber, 5.43 cm
Pressure of Ar in scattering chamber, 1.2×10^{-4} Torr at 300 °K
Attenuation of K beam, 80%

(*a*) How many K atoms per second are scattered out of the beam?
(b) Calculate the cross section (in Å^2).

3.3. Several reactions of alkali metals with halogens have been studied with molecular beam methods [Herschbach and co-workers, *J. Chem. Phys.* **47**, 993 (1967)]. In a typical experiment, a beam of Cs atoms effuses from an oven at 575 °K, with a most probable velocity of 2.68×10^4 cm/sec. This beam intersects a Br_2 beam (most probable velocity 1.84×10^4 cm/sec) at a right angle. The maximum intensity of the CsBr produced in the reaction is found at an angle of 8° with respect to the Cs beam (and toward the motion of the center of mass).

(a) Calculate the center of mass velocity, as well as the velocity of Cs and Br_2 with respect to the center of mass. Draw the corresponding vector diagram.
(*b*) The experiments indicate that the product KBr is scattered in the direction of the Cs atom beam (in CM coordinates). What is the velocity of these product molecules?
(*c*) The average internal energy of Br_2 is about 1.0 kcal/mole. Using this together with $D(KBr) - D(Br_2) = 50.0$ kcal/mole, find the average internal energy of KBr product molecules.

3.4. (From B. H. Mahan.) A wire of radius R and length l has an electric field such that a positive ion *on its surface* has potential energy $V(R) = -eV$. Assume that ions of mass m move with speed g (in a plane perpendicular to the wire) toward the wire.

(*a*) What is b_c, the critical impact parameter for collision of the ion with the surface of the wire?

(b) What is the value of b_c if $\frac{1}{2}mg^2 \ll eV$?

(c) What is the cross section for collisions between the wire and ions of speed g? Remember that the usual spherical symmetry of the target is not present.

3.5. The expression relating the scattering angle ψ in the laboratory coordinate system to that in the center of mass coordinate system (χ) is

$$\tan \psi = \frac{\sin \chi}{\cos \chi + (m_1/m_2)}$$

if the scattering is elastic, and if the velocity of the target (v_2) is much less than that of the incident particle (v_1). This is the case, for example, in the Rutherford α-particle scattering experiment. Derive this relationship. Is the limiting form when $m_1 \ll m_2$ reasonable?

3.6. An approximate value of the cross section for the reaction K + Br$_2$ $\longrightarrow$ KBr + Br is given in the reference cited for Problem 3.3. The angular distribution of scattered K and of KBr is indicated in the figure. Assume that this represents *all* the scattered product. In the same experiment, the attenuation

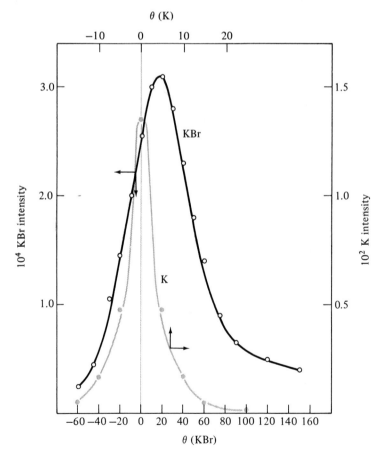

of the K beam is 14 percent. Assume an effective scattering path of 0.45 cm and a Br_2 concentration of 3.3×10^{12} atoms/cc in the scattering region.

(*a*) What is the total cross section?

(*b*) What is the reactive cross section?

3.7. From their trajectory calculations for the reaction $H + H_2$, Karplus et al. find that the reaction probability (also called the opacity) has the form

$$P_r(b, g, v, J) = a \cos \left(\frac{\pi b}{2b_{max}} \right) \qquad b \leq b_{max}$$

$$P_r(b, g, v, J) = 0 \qquad b > b_{max}$$

(*a*) What is the resulting expression for the reactive cross section in terms of b_{max}?

(*b*) For $v = 0$, $J = 0$, $g = 1.18 \times 10^6$ cm/sec, they find $a = 0.39$ and $b_{max} = 0.977$ Å. Calculate the cross section and the corresponding bimolecular rate constant for $H + H_2$.

3.8. Assume that the interaction of two methyl radicals can be represented by the attractive part of the Lennard-Jones potential, with parameters $\sigma_{LJ} = 3.9$ Å and $\epsilon/k = 137$ °K [H. S. Johnston and P. Goldfinger, *J. Chem. Phys.* **37**, 700 (1962)]. If the average relative kinetic energy is $3kT/2$ and the temperature is 400 °K, find the cross section for collision.

4

Rate Constants

for

Elementary Reactions

IN THE PRECEDING CHAPTER we discussed the classical mechanics of bimolecular collisions in the gas phase and derived expressions for rate constants applicable to nonthermal reactions in which the reactants were in well-defined energy states. Usually, however, chemical reactions are studied under conditions of thermal equilibrium, and the reactants have an energy distribution that is consistent with this equilibrium. The goal of this chapter is to convert the nonthermal rate constants previously derived into rate constants for elementary reactions. Again the emphasis will be on bimolecular, gas-phase reactions.

Some statistical mechanical results that are essential to this chapter are contained in Appendix C, which should be referred to before proceeding.

4.1. Reaction Rates for an Ideal Gas

In Appendix B the bimolecular rate constant for collisions of molecules with relative velocity g is shown to be related to the cross section $\sigma(g)$, according to

$$k = \sigma(g)g = \pi b_c^2 g \qquad (B.6)$$

where b_c is the maximum impact parameter for a collisional encounter. The conversion of this expression from a rate constant for a nonthermal reaction to one for an elementary reaction of gases at thermal equilibrium requires only the introduction of a velocity distribution function, $P(g)$, and integration over the entire velocity range. The general form is

$$k = \int_0^\infty \sigma(g)P(g)g \, dg \qquad (4.1)$$

(An equivalent expression could be written in terms of the initial relative kinetic energy in place of the relative velocity.)

The components of Eq. (4.1) are the velocity-dependent reaction cross sections of the preceding chapter and the velocity distribution function of Eq. (C.36). They are illustrated in Fig. 4.1. It is only necessary to modify Eq. (C.36) by replacing the velocity c with the relative velocity g, and by substituting for M the reduced mass

$$\mu = \frac{M_1 M_2}{(M_1 + M_2)}$$

We shall proceed to illustrate the utility of Eq. (4.1) by a few examples.

HARD-SPHERE COLLISIONS

Previously (Section 3.6) we found that the cross section

$$\sigma = \pi(r_1 + r_2)^2$$

was independent of g. Then the rate constant for collisions is

$$k_{coll} = \pi(r_1 + r_2)^2 \int_0^\infty P(g)g \, dg$$

However, the integral is simply the definition of the mean relative velocity

$$\bar{g} = \left(\frac{8kT}{\pi\mu}\right)^{1/2}$$

so that the collisional rate constant is

$$k_{coll} = \left(\frac{8\pi kT}{\mu}\right)^{1/2}(r_1 + r_2)^2 \qquad (4.2)$$

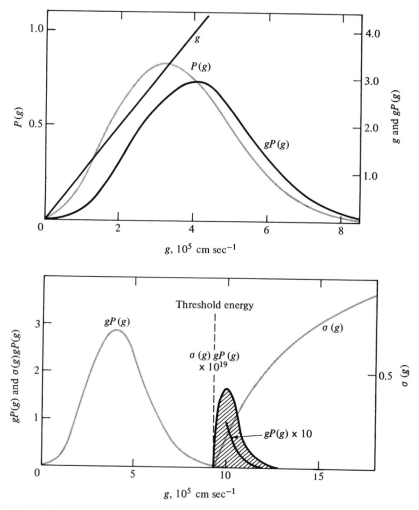

Figure 4.1. *Factors contributing to the rate constant in the expression $k = \int gP(g)\sigma(g)dg$. The cross section is $\sigma(g) = 1.26 \; [1 - (9.30 \times 10^5/g)^2] \; \text{Å}^2$, and $g \, P(g)\sigma(g)dg$ is in units of molecules^{-1} cm^{3}sec^{-1}. The rate constant k is given by the shaded area. The rapid decrease of $gP(g)$ at high velocities leads to very small contributions to the rate constant from reactants with greater than the threshold energy. Therefore, it is difficult to obtain information about the shape of $\sigma(g)$ from thermal reactions.*

With μ in atomic mass units and radii in Å, k_{coll} in units of liters mole^{-1} second^{-1} is

$$k_{coll} = 2.753 \times 10^9 \left(\frac{T}{\mu}\right)^{1/2} (r_1 + r_2)^2$$

The rate (in moles liter^{-1} second^{-1}) at which collisions occur is

$$\mathscr{R} = k_{coll} C_1 C_2$$

Alternatively, the number of collisions per cubic centimeter per second is expressed as

$$Z_{12} = \pi(r_1 + r_2)^2 \bar{g} n_1 n_2 \tag{4.3}$$

where n_1 and n_2 are concentrations in molecules per cubic centimeter, $\bar{g}$ is in centimeters per second, and radii are in centimeters.

If only a single species is present, a factor of one-half must be introduced into Eqs. (4.2) and (4.3) to avoid counting the collisions of each molecule twice. Then the expression corresponding to (4.3) is

$$Z_{11} = 2\pi r^2 n^2 \bar{g}$$
$$= 6.97 \times 10^{32} T^{-3/2} M^{-1/2} P^2 r^2$$

with the molecular weight M in atomic mass units, P in atmospheres, and r in angstrom units. The collisional rate constant is

$$k_{coll} = 7.78 \times 10^9 \left(\frac{T}{M}\right)^{1/2} r^2 \tag{4.4}$$

Another convenient quantity is the collisional frequency ω, which is the number of collisions a single molecule undergoes in one second. This figure is obtained by dividing Eq. (4.3) by n_1. As an indication of the orders of magnitude to be expected, the following numbers pertain to helium at one atmosphere and 300 °K., with a collision radius of 1.1 Å (obtained from the viscosity):

$$\omega_{He} = 6.62 \times 10^9 \text{ collisions sec}^{-1}$$

$$Z_{He, He} = 8.09 \times 10^{28} \text{ collisions cm}^{-3} \text{ sec}^{-1}$$

$$k_{coll} = 8.15 \times 10^{10} M^{-1} \text{ sec}^{-1}$$

HARD-SPHERE COLLISIONS WHEN AN ENERGY BARRIER IS PRESENT

For collisions in which the potential energy has the value V_R when the molecules are separated by a distance $R = r_1 + r_2$, it was shown in Eq. (3.70) that the cross section is

$$\sigma(g) = \pi R^2 \left(1 - \frac{V_R}{E_0}\right) \qquad E_0 \geq V_R$$

and

$$\sigma(g) = 0 \qquad E_0 < V_R$$

The lower limit of the integral in Eq. (4.1) can be moved up to V_R, since there is no contribution to the rate constant from collisions in which the relative kinetic energy is below this value. It is convenient to convert the terms in (4.1) into functions of the relative kinetic energy E_0, which we recall is

$$E_0 = \frac{\mu g^2}{2}$$

The rate constant is

$$k_{coll} = \left(\frac{8\pi}{\mu}\right)^{1/2} \frac{R^2}{(kT)^{3/2}} \int_{V_R}^{\infty} \left(1 - \frac{V_R}{E_0}\right) E_0 e^{-E_0/kT} dE_0$$

The integral contributes

$$(kT)^2 e^{-V_R/kT}\left(1 + \frac{V_R}{kT}\right) - V_R kT e^{-V_R/kT} = (kT)^2 e^{-V_R/kT}$$

Finally,

$$k_{coll} = \left(\frac{8\pi kT}{\mu}\right)^{1/2} R^2 e^{-V_R/kT}$$

$$= \pi R^2 \bar{g} e^{-V_R/kT} \qquad (4.5)$$

Equation (4.5) is the rate constant for hard-sphere collisions multiplied by a Boltzmann factor; it resembles an Arrhenius expression but has an additional (weak) temperature dependence due to the factor $\bar{g}$, the mean relative velocity.

RECOMBINATION OF IONS

Recall that the classical treatment of ion recombination presents a problem because all trajectories are deflected, and the cross section becomes infinitely large. We avoided this problem by defining reactive collisions as those in which the distance of closest approach was some critical value R, and we obtained for the cross section

$$\sigma(E_0) = \pi R^2\left(1 - \frac{z_1 z_2 e^2}{R E_0}\right) \qquad (3.71)$$

The collisional rate constant is given by

$$k_{coll} = \left(\frac{8\pi}{\mu}\right)^{1/2} \frac{R^2}{(kT)^{3/2}} \int_0^{\infty} \left(1 - \frac{z_1 z_2 e^2}{R E_0}\right) E_0 e^{-E_0/kT} dE_0$$

Subsequent development is parallel to that in the preceding example, except that the lower limit of integration is zero because $\sigma(E_0)$ is finite for all values of E_0 (the second term in the cross section expression is positive). After evaluation of the integral, we obtain

$$k_{coll} = \left(\frac{8\pi kT}{\mu}\right)^{1/2} R^2 \left(1 - \frac{z_1 z_2 e^2}{RkT}\right)$$

$$= R^2 \bar{g} \left(1 + \frac{|z_1 z_2| e^2}{RkT}\right) \tag{4.6}$$

The term in parentheses has a value of 280 at $300°K$ for singly charged ions approaching to a critical distance of 2 Å; hence the rate constant is much larger than that for hard-sphere collisions.

ION-MOLECULE REACTIONS

Previously we showed that the potential energy for the interaction of an ion and an induced dipole is

$$V(r) = \frac{-z_1^2 e^2 \alpha_2}{2r^4}$$

and that the resulting cross section is given by

$$\sigma(g) = 2\pi \left(\frac{z_1^2 e^2 \alpha_2}{\mu g^2}\right)^{1/2} \tag{3.77}$$

Substituting this equation into Eq. (4.1) gives a rate constant

$$k = 2\pi z_1 e \left(\frac{\alpha_2}{\mu}\right)^{1/2} \int_0^\infty P(g)\, dg$$

Since the integral is unity by definition, the rate constant is

$$k = 2\pi z_1 e \left(\frac{\alpha_2}{\mu}\right)^{1/2} \tag{4.7}$$

4.2. *Bimolecular Reactions Without Activation Energy:* *Comparison of Theory and Experiment*

The preceding section provides methods of calculating rate constants for some particular types of reaction. A comparison of calculated values with those determined experimentally will provide some measure of the validity of the simple models used thus far.

A large number of reactions with no activation energy have been discussed by Johnston and Goldfinger (2). These reactions comprise the recombination of polyatomic radicals as well as the addition of atoms to stable molecules. Some of their results are given in Table 4.1, where experimental values are compared with values calculated from Eqs. (4.2) or (4.4). The tabulated statistical factor accounts for the electronic multiplicity of the reactants and product (2 for polyatomic radicals, 4 for Cl, 1 for stable molecules). In other words, it is assumed that the combining radicals must have antiparallel electron spins if they are to produce a stable product. The small range of $r_1 + r_2$ and of μ leads to rate constants that differ by, at most, a factor of 2. The more complicated Lennard-Jones expression gives slightly larger rate constants (2), but again the range is narrow.

Table 4.1. *Rate Constants for Bimolecular Reactions without Activation Energy (at 400 °K)*

Reaction	$r_1 + r_2$, Å[a]	Statistical factor	log k (M^{-1} sec^{-1}) Calculated	Observed
$CH_3 + CH_3 \rightarrow C_2H_6$	3.9	$\frac{1}{4}$	10.58	10.3[b]
				10.5[c]
$CF_3 + CF_3 \rightarrow C_2F_6$	4.0	$\frac{1}{4}$	10.27	10.4[d]
$CCl_3 + CCl_3 \rightarrow C_2Cl_6$	5.4	$\frac{1}{4}$	10.42	10.9[e]
				9.6[f]
$C_2H_5 + C_2H_5 \rightarrow C_4H_{10}$	4.4	$\frac{1}{4}$	10.54	10.5[g]
$CH_3 + Cl \rightarrow CH_3Cl$	3.6	$\frac{1}{4}$	10.44	11.6[h]
$C_2Cl_4 + Cl \rightarrow C_2Cl_5$	4.2	$\frac{1}{2}$	10.95	9.6[h]

[a]Reference 2.
[b]A. Shepp, *J. Chem. Phys.* **24**, 939 (1956); R. E. March and J. C. Polanyi, *Proc. Royal Soc. (London)* **A273**, 360 (1963).
[c]R. Gomer and G. B. Kistiakowsky, *J. Chem. Phys.* **19**, 85 (1951).
[d]P. B. Ayscough, *ibid.* **24**, 944 (1956).
[e]J. M. Tedder and J. C. Walton, *Trans. Faraday Soc.* **63**, 2464 (1967).
[f]G. R. de Maré and G. H. Huybrechts, *Chem. Phys. Letters* **1**, 64 (1967).
[g]A. Shepp and K. O. Kutschke, *J. Chem. Phys.* **26**, 1020 (1957).
[h]R. Eckling, P. Goldfinger, G. Huybrechts, G. Martens, L. Meyers, and S. Smoes, *Chem. Ber.* **93**, 3014 (1960).

Second-order rates for radical recombinations have been measured using the rotating-sector technique (Appendix A), and in all cases cited, the effect of temperature on the rate constant is small.

In addition to the absolute rate constants of Table 4.1, a large number of relative recombination rates have been determined. When two different radicals, R_1 and R_2, are present simultaneously, they can recombine according to

$$R_1 + R_1 \xrightarrow{k_{11}} R_1R_1$$

$$R_1 + R_2 \xrightarrow{k_{12}} R_1R_2$$

$$R_2 + R_2 \xrightarrow{k_{22}} R_2R_2$$

A product analysis then yields the ratio $k_{12}^2/k_{11}k_{22}$, which would be 4 if it were determined by the statistical factor alone. It is close to this value for a large number of alkyl radicals; thus the individual rate constants must be nearly equal. The simple collision model (either hard sphere or Lennard-Jones) is in accord with this conclusion.

Note that the recombination of atoms to form diatomic molecules has not been considered here. The reason is that such reactions are inevitably termolecular, involving the presence of a third body to absorb the energy released when the new bond is formed. Two types of mechanism are possible:

1. Energy-Transfer Model

$$R + R \rightleftharpoons RR^*$$

$$RR^* + M \longrightarrow RR + M$$

(where the asterisk denoted a vibrationally excited species)

2. Bound-Complex Model

$$R + M \rightleftharpoons RM^*$$

$$RM^* + M \rightleftharpoons RM + M \text{ or } RM^* + R \longrightarrow RR + M$$

$$RM + R \longrightarrow RR + M$$

A good discussion of these processes (and their inverse, the unimolecular dissociation of diatomic molecules) is given by Bunker (3).

Of course, the difference in behavior between the recombination of atoms and the recombination of polyatomic radicals is quantitative rather than qualitative. The discussion of unimolecular reactions in Chapter 5 indicates that the lifetime of a vibrationally excited molecule depends on the number of atoms it contains. Therefore an excited polyatomic molecule is stable enough so that energy transfer is generally not a rate-determining step. At very low pressures, this fact will no longer be true; neglect of this complication led to some errors in the early interpretation of recombination rate measurements.

Rate constants for reactions of atoms with stable molecules or radicals cover a much wider range than do those for radical recombinations (Ref. 2 contains a list of 14 such reactions). The simple collision theory does not adequately predict this variation.

The reaction of ions with molecules provides another class of reactions to which a collision model can be usefully applied. Such reactions were first studied in mass spectrometers, where it was observed that certain secondary ion currents depended on the square of the gas pressure in the source. With hydrogen in the mass spectrometer, for example, the dependence could be explained by the occurrence of the reactions

$$H_2 + e^- \longrightarrow H_2^+ + 2e^- \quad \text{(primary ion formation)}$$

$$H_2^+ + H_2 \longrightarrow H_3^+ + H \quad \text{(secondary ion formation)}$$

The intensity of the H_2^+ beam will depend on P_{H_2}, whereas that of the H_3^+ ion beam will be proportional to $P_{H_2}^2$; hence the ratio $I(H_3^+)/I(H_2^+)$ depends on P_{H_2}. Actual determination of the bimolecular rate constant requires a knowledge of the electric fields within the mass spectrometer source, as well as the measurement of the very low pressures in this region. More recently, experiments have been done in which a beam of primary ions, produced in one mass spectrometer, interacts with the neutral target molecules in a scattering chamber and the secondary ion beam is examined with a second mass spectrometer (see. Section 3.11).

A large number of observed rate constants, some of which are given in Table 4.2, are found to agree very well with the predictions of Eq. (4.7). An interesting feature of this expression is that unlike most of the collisional rate-constant formulas, it does not contain a "critical distance," which is always difficult to define. The predicted temperature independence of the rate constant is verified experimentally.

There are also many ion-molecule reactions with rate constants that are orders of magnitude smaller than those predicted. Such a situation does not necessarily indicate a breakdown of the model, which only gives a rate of collision. Subsequently, the collision complex may break up in different ways to form various products; if some of these products are not observed, the rate constant determined from secondary ion intensities will be too low.

Table 4.2. *Rate Constants for Exothermic,*
Bimolecular Reactions of Ions with Molecules

Reactants	$10^{24}\alpha$, cm^{3a}	$\log k$ (M^{-1} sec^{-1}) Calculated	Observed
H_2^+, H_2	0.79	12.10	12.10[b]
D_2^+, D_2	0.79	11.95	11.94[b]
Ar^+, H_2	0.79	11.95	12.00[b]
O^-, H_2	0.79	11.98	11.78[b]
H_2^+, O_2	1.60	12.11	12.70[b]
CH_4^+, CH_4	2.60	11.90	11.90[c]

[a]From J. O. Hirschfelder, C. F. Curtis, and R. B. Bird, *Molecular Theory of Gases and Liquids*, John Wiley & Sons, New York, 1954, p. 950.

[b]From an extensive tabulation in V. Cermak, A. Dalgarno, E. E. Ferguson, L. Friedman, and E. W. McDaniel, *Ion-Molecule Reactions*, John Wiley & Sons, New York, 1970.

[c]A. Giardini-Guidoni and L. Friedman, *J. Chem. Phys.* **45**, 937 (1966).

Additional information about ion-molecule reactions may be found in recent review articles (4).

Ion recombination rates can also be treated by a collision model, by use of Eq. (4.6). The only recent determinations of these rate constants have been done by Mahan and his students (5). Photoionization of a nitric oxide-nitrogen dioxide mixture produces NO^+ ions and electrons, the latter forming NO_2^- ions by a subsequent process. The ultraviolet light source is then abruptly shut off, and after a suitable time, during which ions recombine (1 to 50 milliseconds), a high voltage is applied to collecting plates in the reaction chamber. The resulting ion current is measured with an oscilloscope and can be related to the number of ions remaining at the time of measurement. The rate is found to be proportional to the pressure of added inert gas, but there is a finite value at zero pressure. This fact indicates that two processes are occurring simultaneously, one of them involving three-body interactions just as the recombination of atoms does. The other process (independent of pressure) is the one of interest to us, for it is bimolecular. Rate constants for the following combinations have been determined:

$NO^+ + NO_2^-$

$NO^+ + SF_6^-$

$C_6H_6^+ + SF_6^-$

$I_2^+ + I^-$ (or I_2^-)

The rate constants are all extremely large, in the range of 4 to 15×10^{13} $M^{-1} \sec^{-1}$. These values cannot be compared with values predicted from Eq. (4.6) without knowledge of R, the distance at which electron transfer takes place. However, we can work backward to calculate R from this equation and the measured rate constants. In all the reactions studied, this distance is very close to 20 Å. Mahan has suggested that this large distance corresponds to the point at which the coulomb potential curve for A^+, B^- coincides with energy levels where A is in the ground state and B is in a Rydberg state (an electronic state in which one electron is in an orbital that resembles an atomic orbital with a large principal quantum number).

It is absolutely necessary to take into account the coulombic interaction between ions; a simple hard-sphere calculation with the usual collision diameters would lead to results that are low by factors of one hundred or more.

4.3. Failure of the Collision Model for Bimolecular Reactions

In Section 4.2 we have seen a number of successful applications of simple collision models, and a few failures. One might hope that bimolecular reactions with activation energies could be handled in the same way, provided one knew the activation energy to use as V_R in Eq. (4.5). The pre-exponential factor in the rate constant should then be given by a hard-sphere collision number, and the rate constant would be of the form

$$k_{\text{bim}} = k_{\text{coll}} e^{-E_a/kT} \tag{4.8}$$

However, it was realized very early in the history of chemical kinetics that most pre-exponential factors were much smaller. For example, in the hydrogen-bromine reaction, the step $Br + H_2 \rightarrow HBr + H$ has a pre-exponential factor about one-tenth the collision number, and in the reverse reaction this factor drops to 0.03.

The first attempt to rationalize these observations was the introduction of a "steric factor" p, so that Eq. (4.8) becomes

$$k_{\text{bim}} = p k_{\text{coll}} e^{-E_a/kT} \tag{4.9}$$

The idea behind this rationale is that the collisional rate only measures the frequency with which molecules approach to within a certain distance; it does not allow for the possibility that the orientation of reactant molecules, as well as their separation, is important. The idea is perfectly reasonable, but unfortunately there is no a priori method for predicting p, for example, on the basis of molecular geometries.

What else might contribute to the failure of a collision model? The types of potential energy expression we have used are probably too simple. In fact, the discussion of the $H + H_2$ reaction in the preceding chapter described a way of taking this factor into account. In order to obtain the rate constant for this elementary reaction, it is only necessary to combine the numerical values of the cross section as calculated by Karplus et al., with a thermal velocity distribution.

Even if a simple potential energy expression is correct, there remains the problem of defining the intermolecular distance at which reaction occurs.

Finally, and to some extent this point is involved in the preceding remarks, we have treated molecules as structureless "blobs," whereas, in fact, their rotational and vibrational degrees of freedom should be taken into account in any reasonable model.

If the internal states (vibrational and rotational) of reactants and products are specified by i, j, k, and l, the expression for a bimolecular reaction may be written

$$A(i) + B(j) \longrightarrow P(k) + R(l)$$

The total reactive cross section is written $\sigma_{ij}^{kl}(g)$ to indicate that it depends on internal states as well as on the relative velocity g. The overall rate constant from Eq. (4.1) is then

$$k = \sum_{i,j,k,l} P_{A_i} P_{B_j} \int_0^\infty \sigma_{ij}^{kl}(g) P(g) g \, dg \tag{4.10}$$

where P_{A_i} and P_{B_j} are the fractions of A and B molecules in states i and j. The difficulty in using this equation arises in the evaluation of cross sections for individual internal states of reactants and products. (Although, in fact, such evaluation is just what was done in the trajectory calculations for the $H + H_2$ reaction described in Section 3.9.) For this reason, Eq. (4.10), although formally correct, is not very useful.

4.4. The Activated Complex Theory of Bimolecular Rate Constants

Instead of calculating many trajectories, and then deriving cross sections from them, one would like an approach that combines statistical mechanics with the potential energy surface for the reacting system (6). Attempts in this direction led to the development, in the 1930s, of the activated complex theory of rate constants (also called the transition state theory and the absolute rate theory.)*

In Section C.4 it is pointed out that the state of a molecule, or a reacting system, could be defined by a representative point in phase space. This phase space has six dimensions for each of the r atoms in the system, for its three coordinates and three conjugate momenta. The state is defined by specifying the position of a volume element

$$d\tau = \frac{dq_1 \, dq_2 \, \cdots \, dp_{3r}}{h^{3r}}$$

Clearly, we can distinguish two general regions in phase space, one corresponding to reactants and the other to products. A boundary surface, dividing these two regions, is placed at the saddlepoint of the potential energy surface. This configuration is called the activated complex or the transition state, and this specification defines a unique reaction coordinate $q^\ddagger$ such that the total energy can be divided into two parts

$$\epsilon = \epsilon^\ddagger(q^\ddagger, p^\ddagger) + \epsilon'(q_2, p_2, \ldots, q_{3r}, p_{3r}) \tag{4.11}$$

(This separation can always be performed if the coordinate $q^\ddagger$ is orthogonal to all other coordinates $q_2, \ldots, q_{3r}$.) The number of systems in a state represented by a point within $d\tau$ will be $dN(q^\ddagger, p^\ddagger, q_2, \ldots, p_{3r})$.

---

*The names of Wigner, Pelzer, M. Polanyi, Evans, and particularly Eyring are associated with this development. See Glasstone, Laidler, and Eyring, *The Theory of Rate Processes*, McGraw-Hill Book Co., New York, 1941.

With appropriate units, this number can be replaced by a concentration, $C^\ddagger(q^\ddagger, p^\ddagger, q_2, \ldots, p_{3r})$. The time required for passage through the boundary surface will be designated $t^\ddagger(q^\ddagger, p^\ddagger)$; it is assumed to depend only on the reaction coordinate and its conjugate momentum.

Each volume element of the phase space of the activated complex contributes an amount to the rate of reaction

$$d\mathcal{R}_{AB} = \frac{C^\ddagger(q^\ddagger, p^\ddagger, \ldots, p_{3r})\kappa}{t^\ddagger(q^\ddagger, p^\ddagger)} \tag{4.12}$$

where κ is a factor introduced to account for those systems that cross the boundary but return to the reactant side. This factor is expected to be a function of the particular state, but the actual dependence is not usually known, and κ will later be assumed to be unity for all activated complex states.

A function, similar to a distribution function, may be defined as

$$P \, d\tau = \frac{C^\ddagger(q^\ddagger, p^\ddagger, \ldots, p_{3r})}{C_A C_B}$$

where C_A and C_B are the reactant concentrations; Eq. (4.12) becomes

$$d\mathcal{R}_{AB} = \frac{\kappa P \, d\tau}{t^\ddagger} C_A C_B \tag{4.13}$$

Integration over all phase space now gives the rate

$$\mathcal{R}_{AB} = \int \left(\frac{\kappa P \, d\tau}{t^\ddagger} \right) C_A C_B$$

and by comparison with the rate of a bimolecular reaction, we have for the rate constant

$$k = \int \frac{\kappa P \, d\tau}{t^\ddagger} \tag{4.14}$$

At this point in the development, the assumption is made that thermal equilibrium exists among all states in the phase space region on the reactant side of the boundary surface, as well as in the activated complex. Consequently, it is possible to write $P \, d\tau$ in terms of a Boltzmann factor for the activated complex and partition functions for the reactants [sec. Eq. (C. 38)]. This leads to

$$P \, d\tau = \left(\frac{e^{-\epsilon^\ddagger/kT} dq^\ddagger \, dp^\ddagger}{h} \right) \frac{(e^{-\epsilon'/kT} dq_2 \, \cdots \, dp_{3r}/h^{3r-1})}{Q_A Q_B}$$

The rate constant of Eq. (4.14) becomes

$$k = \left[\frac{1}{h} \int\!\!\int \frac{\kappa}{t^\ddagger} e^{-\epsilon^\ddagger/kT} dq^\ddagger \, dp^\ddagger \right]\left[\frac{1}{h^{3r-1} Q_A Q_B} \int \cdots \int e^{-\epsilon'/kT} dq_2 \, \cdots \, dp_{3r} \right] \tag{4.15}$$

The multiple integral in the second bracketed expression has the form of a classical partition function for $3r - 1$ degrees of freedom. It is convenient to use the potential energy at the saddlepoint as the energy origin of the activated complex; origins for the reactants are their respective potential energy minima (Fig. 4.2). The difference between these energies is the barrier height V_a. The second term in brackets of Eq. (4.15) then becomes

$$\frac{Q^{\ddagger}_{3r-1}e^{-V_a/kT}}{Q_A Q_B} \tag{4.16}$$

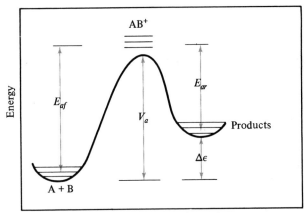

Figure 4.2. *Schematic representation of the energy relation between reactants, activated complex, and products. The activation energy for the forward reaction, E_{af}, includes V_a and the difference in internal thermal energies of $A + B$ and $AB^{\ddagger}$.*

In evaluating the integral in the first set of brackets, only motion across the boundary surface in the direction from reactants to products is to be included. This condition is taken care of by integrating only over positive values of $p^{\ddagger}$. It is assumed that κ is unity. The limits on $q^{\ddagger}$ are zero and $l^{\ddagger}$, which is the distance along the reaction coordinate traversed in time $t^{\ddagger}$. Hence the integral to be evaluated is

$$h^{-1} \int_0^{l^{\ddagger}} \int_0^{\infty} \left(\frac{1}{t^{\ddagger}}\right) e^{-\epsilon^{\ddagger}/kT} dq^{\ddagger} \, dp^{\ddagger}$$

In a sufficiently small region at the top of the barrier, the potential energy $V(q^{\ddagger})$ is constant and has already been included in the factor $\exp(-V_a/kT)$. Therefore $\epsilon^{\ddagger}$ is only the kinetic energy, $m^{\ddagger}(\dot{q}^{\ddagger})^2/2$. The element of phase space appearing in the foregoing integral is

$$dp^{\ddagger} \, dq^{\ddagger} = (m^{\ddagger} \, d\dot{q}^{\ddagger}) \, dq^{\ddagger}$$

$$= (m^{\ddagger} \, \dot{q}^{\ddagger} \, d\dot{q}^{\ddagger}) \frac{dq^{\ddagger}}{\dot{q}^{\ddagger}} = d\epsilon^{\ddagger} \, dt$$

The integral becomes

$$h^{-1} \int_0^{\infty} \int_0^{t^{\ddagger}} \left(\frac{1}{t^{\ddagger}} \right) e^{-\epsilon^{\ddagger}/kT} d\epsilon^{\ddagger} \, dt$$

$$= h^{-1} \int_0^{\infty} e^{-\epsilon^{\ddagger}/kT} d\epsilon^{\ddagger} = \frac{kT}{h} \tag{4.17}$$

This factor kT/h may be interpreted as the reciprocal of the average "lifetime" of the activated complex, or as the average frequency of passage over the barrier.*

Combining Eqs. (4.15), (4.16), and (4.17), we finally obtain

$$k = \frac{kT}{h} \frac{Q^{\ddagger}_{3r-1}}{Q_A Q_B} e^{-V_a/RT} \tag{4.18}$$

(We have expressed V_a in kilocalories per mole and replaced k by R in kilocalories per mole per degree.) Although this expression has been derived for (and is usually applied to) bimolecular reactions, it can be modified in an obvious way for unimolecular or termolecular reactions.

Two of the major assumptions in the activated complex theory of rate constants are

1. Reactant molecules are distributed among their possible internal energy states as they would be at thermal equilibrium;

2. Activated complexes are similarly distributed among their possible states.

For a special class of reactions, termed *vibrationally adiabatic*, it may be shown that Assumption 2 follows from the less-restrictive Assumption 1 (7). In a vibrationally adiabatic reaction, the reactant vibrations (except the one that becomes the reaction coordinate) evolve smoothly into those of the activated complex, and finally into those of the product, without any change in vibrational quantum numbers. (The *amount* of vibrational energy is not constant, in general, since the vibrational

---

*A rationalization for the order of magnitude of this quantity can be derived from the Uncertainty Principle,

$$\Delta\epsilon \, \Delta t \geq h$$

where Δt is the lifetime of a state and $\Delta\epsilon$ is the precision with which the energy of the state can be specified. For the activated complex with "lifetime" $t^{\ddagger}$, $\Delta\epsilon$ must include at least the thermal distribution, so that $\Delta\epsilon^{\ddagger}$ is of the order of kT. Since the state is unstable (at a potential maximum), the equality sign is appropriate and

$$\frac{1}{t^{\ddagger}} \sim \frac{kT}{h}$$

frequencies of the three species differ.) Trajectory calculations for the $H + H_2$ reaction indicate that this adiabatic condition is met, at moderate energies.

4.5. *Detailed Formulation of the Rate Constant*

If each of the partition functions in the preceding expression can be factored into components for the various forms of energy, Eq. (4.18) can be written out in more detail as

$$k = \frac{kT}{h}\left(\frac{Q^{\ddagger}}{Q_A Q_B}\right)_{\text{trans}} \left(\frac{Q^{\ddagger}}{Q_A Q_B}\right)_{\text{rot}} \left(\frac{Q^{\ddagger}}{Q_A Q_B}\right)_{\text{vib}} \left(\frac{Q^{\ddagger}}{Q_A Q_B}\right)_{\text{el}} e^{-V_a/RT} \qquad (4.19)$$

Each factor in the above expression will be discussed separately.

FREQUENCY FACTOR

The factor kT/h has a numerical value of $2.08 \times 10^{10}\, T$ (sec^{-1}), or about $6 \times 10^{12}\, \text{sec}^{-1}$ at 300 °K. This factor is the same for all reactions and is appropriate when motion through the pass on the potential energy surface can be treated by the method of the preceding section.

Unfortunately, use of the method described is not always possible. Quantum mechanics shows that every particle has certain wave properties and that the de Broglie wavelength associated with a particle is

$$\lambda = \frac{h}{p}$$

The momentum of a macroscopic object is always so large that λ is very much shorter than the object itself. But the wave nature of atomic-sized particles becomes extremely important when they are subject to a potential energy that changes sharply in a distance equal to the particle wavelength. This effect is attested to by such phenomena as electron diffraction by molecules and crystal lattices. For calibration, note that a hydrogen atom at 300 °K has a de Broglie wavelength of 1 Å.

One of the consequences of this wave nature is that a particle confined by a potential barrier (Fig. 4.3) has a finite probability of leaking through the barrier even when its energy is less than the barrier height V_0. This phenomenon is called the tunnel effect and, of course, is unimportant for macroscopic particles. There is also a finite probability for reflection from the barrier, even when the particle energy is greater than the barrier height. If the reaction coordinate involves the motion of a light atom (a hydrogen atom or proton), then tunneling may become an important path for the reaction to follow.

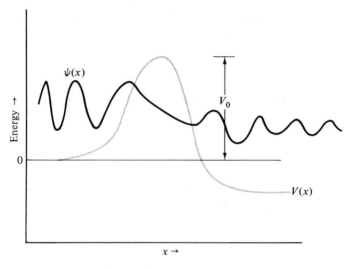

Figure 4.3. *Penetration of a potential barrier,* $V(x)$, *by a particle with wave function* $\psi(x)$. [*After Bunker.*]

Methods of calculating the extent of tunneling for special barrier shapes are described elsewhere (8). The result is a correction factor Γ_t, by which the classical rate-constant expression must be multiplied to take tunneling into account.

TRANSLATIONAL PARTITION-FUNCTION RATIO

There is obviously no problem in evaluating this term; since the mass of the activated complex is $M^{\ddagger} = M_A + M_B$, the ratio is

$$\left(\frac{M_A + M_B}{M_A M_B}\right)^{3/2}\left(\frac{h^2}{2\pi kT}\right)^{3/2} = \frac{h^3}{(2\pi \mu_{AB}kT)^{3/2}}$$

At 300 °K this quantity will typically be about 2×10^{-3}, in units of moles per liter.

ROTATIONAL PARTITION-FUNCTION RATIO

The rotational partition function of the activated complex can be evaluated from the atomic masses and the values of all internuclear distances at the saddlepoint. A statistical term is generally included here to account for the multiplicity of reaction paths. For example, in a comparison of rates of hydrogen atom abstraction from various hydrocarbons

$$X + HR \longrightarrow XH + R$$

it is necessary to consider the number of equivalent hydrogen atoms available in each reactant: one for CCl_3H, four for CH_4, and so on. This point has been much discussed (9), but the simplest approach is provided by the following prescription.

Omit all the symmetry numbers that would ordinarily be included in the individual rotational partition functions of reactants and activated complex. Label all identical atoms in the reactants, and count the number of different activated complexes that can be formed by permuting the labels. Multiply the rate constant by this number, $s^{\ddagger}$. The preceding example of methane would give $X \ldots H_a \ldots CH_b H_c H_d$, $X \ldots H_b \ldots$ $CH_a \ H_c H_d$, etc., and a factor of four for $s^{\ddagger}$.

In making a rough estimate of the rate constant, it may be assumed that each rotational degree of freedom (for reactant or activated complex) will contribute a factor of 10 to 100. The number of degrees of rotational freedom remaining in the partition function ratio depends on whether the reactants and transition state are linear or not.

VIBRATIONAL PARTITION-FUNCTION RATIO

The vibrational partition function for the activated complex has one less term than the corresponding normal molecule would have; that is, there are $3r - 7$ factors if it is nonlinear, $3r - 6$ if it is linear. The reason is that the motion along $q^{\ddagger}$ has been treated as a one-dimensional translation. Nonetheless, the usual classical equations of motion for the vibrations can be set up and solved for vibrational frequencies, provided the necessary force constants are known. These constants depend on the curvatures of the potential energy surface at the saddle point (Section 2.6). When the vibrational secular equation is set up and solved, there will be one zero or imaginary frequency, resulting from the upside-down nature of the potential energy curve along the reaction coordinate. This frequency is *not* the frequency of passage over the barrier, although this point is often confused in derivations of activated complex theory. In most application of activated complex theory the imaginary frequency is ignored.

In deriving Eq. (4.19) we began with a classical statistical mechanical expression, so it must be corrected for possible quantum effects. As stated earlier, translational and rotational partition functions are essentially classical at ordinary temperatures. The quantum correction for motion along the reaction coordinate has already been discussed. The only other source of quantum effects is in the vibrational partition functions, and all the Q_{vib} terms of Eq. (4.19) should be in the form for a quantized oscillator,

$$Q_{\text{vib}} = \Pi_i e^{-u_i/2}(1 - e^{-u_i})^{-1}$$

In making rate constant estimates, it is frequently assumed that all the $(1 - e^{-u})$ factors are unity. This assumption is approximately correct for the vibrations of normal molecules at ordinary temperatures. If large parts of the activated complex resemble normal molecules in their vibrational properties, this approximation will be a good one for many of the vibrational Q's of the activated complex. For vibrations involving bonds broken or formed in the reaction, however, it is *not* a good approximation, because such bonds are weak and the associated vibrations will be of unusually low frequency. Their contributions to $Q_{\text{vib}}^{\ddagger}$ may be much closer to the classical value $\mathbf{k}T/h\nu$ than to unity. A concise and direct way of estimating such contributions is described by Johnston (10), but it still requires a knowledge of force constants for reactants and activated complex.

ELECTRONIC CONTRIBUTIONS

The last two terms in (4.19) depend on electronic properties. The electronic partition-function ratio is determined by the degeneracy factors of the electronic ground states; different rate constants would be anticipated for higher electronic states. The methods of estimating V_a have been discussed in Chapter 2.

4.6. Bimolecular Rate Constants: Arrhenius Parameters

A direct comparison between a rate constant measured at a single temperature and the corresponding constant calculated from one of the preceding models is not very enlightening. The reason is that the Boltzmann factor $\exp(-V_a/RT)$ or $\exp(-V_R/RT)$ is so important, and it depends only on the potential energy of interaction. Thus the rate constant comparison becomes a test of quantum mechanics and not of chemical kinetic theory. It is much better to compare the pre-exponential factor and the activation energy of the Arrhenius expression separately.

How can this step be done when the theoretical rate constant expressions, such as Eqs. (4.5) and (4.19), do not have the same form as the Arrhenius rate expression, $k = Ae^{-E_a/RT}$? A general form of the theoretical equation is

$$k = B(T)e^{-V_a/RT} \qquad (4.20)$$

The definition of the activation energy in terms of experimental observables (Section 1.6) is

$$E_a = \frac{-R \, d \ln k}{d(1/T)}$$

If we define a dimensionless quantity Θ by

$$\Theta = \frac{d \ln B(T)}{d \ln T} = \frac{-d \ln B(T)}{T \, d(1/T)} \tag{4.21}$$

then

$$E_a = V_a + RT\Theta \tag{4.22}$$

Thus the activation energy is not entirely independent of temperature.

B(T) or Θ can be factored into a classical part and a quantum correction (for the vibrations and reaction coordinate). The contribution of each degree of freedom to the classical part of Θ can be found from the partition function expressions, and is as follows:

1. Each translational degree of freedom contributes $\frac{1}{2}$.
2. Each rotational degree of freedom contributes $\frac{1}{2}$.
3. Each classical vibration contributes 1.
4. The classical reaction coordinate contributes 1.

If one applies these rules to the various cases of atom + atom, atom + diatomic molecule, etc., one finds in general that

$$\Theta_{\text{classical}} = \frac{1}{2}(f^{\ddagger} - f_{\text{reactants}} - \chi)$$

where $\chi = 1$ for a reaction of an atom with a diatomic molecule via a nonlinear activated complex and $\chi = 0$ for all other cases. In this expression, f indicates the number of internal degrees of freedom, including the reaction coordinate.

For each vibration, the additional quantum contribution is

$$\frac{-d \ln \Gamma}{T \, d(1/T)} = \frac{-d \ln \left[(u/2)/\sinh(u/2) \right]}{T \, d(1/T)}$$

$$= \left(\frac{u}{2} \right) \coth \left(\frac{u}{2} \right) - 1 \tag{4.23}$$

When tunneling must also be taken into account, the temperature variation of the tunneling correction must be included by a term

$$\theta_t = \frac{d \ln \Gamma_t}{d \ln T}$$

This expression is usually computed numerically, and since tunneling becomes less important as the temperature increases, it is negative. Combining all the preceding terms, we have

$$\Theta = \frac{1}{2}(f^{\ddagger} - f_{\text{reactants}} - \chi) + \sum_{i=1}^{3r-7} \left[\left(\frac{u_i^{\ddagger}}{2} \right) \coth \left(\frac{u_i^{\ddagger}}{2} \right) - 1 \right]$$

$$- \sum_{\text{reactants}} \sum_{i=1}^{3r-6} \left[\left(\frac{u_i}{2} \right) \coth \left(\frac{u_i}{2} \right) - 1 \right] + \theta_t \tag{4.24}$$

In this expression, the summation over $3r - 7$ vibrational frequencies of the activated complex is to be replaced by a summation over $3r - 6$ frequencies if the complex is linear. Similarly, a linear reactant has $3r - 5$ instead of $3r - 6$ vibrational frequencies.

The Arrhenius pre-exponential factor is

$$A = ke^{E_a/RT} = [B(T)e^{-V_a/RT}][e^{V_a/RT}e^{\ominus}]$$
$$= B(T)e^{\ominus} \tag{4.25}$$

Equation (4.25) predicts that A is not temperature independent; nevertheless, the variation in A over an experimentally attainable range of temperatures is small compared with the change in $\exp(-V_a/RT)$. If this were not so, one could hope to choose between various models for the reaction process (e.g., activated complex vs. collision models) on the basis of the experimental form of $B(T)$.

Once the proper expression for $\ominus$ has been obtained, it is relatively easy to find the activation energy and the pre-exponential factor, but the detailed general expressions are cumbersome and will not be written out.

4.7. Example 1: Hard-Sphere Collisions

The activated complex theory can be applied to this case and the result compared with the rate constant expression obtained from the classical trajectory approach.

For the collision of two structureless particles, the activated complex is simply the two particles separated by a distance R. The reaction coordinate is the line connecting their centers, and there are no vibrational degrees of freedom. The moment of inertia is μR^2, and the rotational partition function is

$$(Q^{\ddagger})^2_{\text{rot}} = \frac{8\pi^2 \mathbf{k}T}{sh^2}\mu R^2$$

where $s = 2$ if $A = B$, and $s = 1$ if $A \neq B$. The rate constant according to Eq. (4.19) is

$$k = \frac{\mathbf{k}T}{h}\frac{h^3}{(2\pi\mu\mathbf{k}T)^{3/2}}\frac{8\pi^2\mathbf{k}T}{sh^2}\mu R^2\left(\frac{Q^{\ddagger}}{Q_A Q_B}\right)_{\text{el}}e^{-V_R/RT}$$
$$= s^{-1}\left(\frac{8\pi\mathbf{k}T}{\mu}\right)^{1/2}R^2\left(\frac{Q^{\ddagger}}{Q_A Q_B}\right)_{\text{el}}e^{-V_R/RT}$$

This expression is similar to Eq. (4.5), which was found by a purely classical calculation based on the detailed mechanics of the two-body collision.

The Arrhenius parameters are the pre-exponential factor

$$A = B(T)e^{\Theta} = s^{-1}\left(\frac{8\pi \mathbf{k}T}{\mu}\right)^{1/2} R^2 \left(\frac{Q^{\ddagger}}{Q_A Q_B}\right)_{el} e^{1/2}$$

and the activation energy

$$E_a = V_R + \tfrac{1}{2} RT$$

4.8. *Example 2: The* H + H₂ *Reaction*

This reaction has always been of fundamental importance to chemical kinetics because it is one of the few examples where quantum mechanics has been adequately applied to the potential energy problem. It also provides an example of a large class of reactions in which a hydrogen atom is abstracted by either an atom or a radical.

For the reaction of an atom with a diatomic molecule, by way of a linear activated complex, Eq. (4.19) gives for the rate constant

$$k = \frac{s^{\ddagger}\Gamma_t h^2}{(2\pi)^{3/2}(\mathbf{k}T)^{1/2}}\left(\frac{Q_{H_3}}{Q_H Q_{H_2}}\right)_{el}\left(\frac{3m_H}{2m_H^2}\right)^{3/2}_{trans}\left(\frac{2m_H[R_{HH}^{\ddagger}]^2}{\tfrac{1}{2}m_H[R_{HH}]^2}\right)_{rot}$$
$$\times \left(\frac{Q^{\ddagger}_{stretch}Q^{\ddagger 2}_{bend}}{Q_{H_2}}\right)_{vib} e^{-V_a/RT}$$

For the necessary properties of the activated complex, we use values recently calculated by Shavitt (11). He finds it to be linear and symmetrical, with an H—H distance of 0.934 Å. Vibrational frequencies are 2025 cm⁻¹ (stretch) and 973 cm⁻¹ (degenerate bend). The hydrogen molecule has a bond length of 0.742 Å and a vibrational frequency of 4395 cm⁻¹.

If it is assumed that the multiplicity of the ground electronic states of H and H₃ are the same, the electronic factor is unity. The reaction-path multiplicity contributes a factor of two. To a fairly good approximation, all values of u are large enough so that the vibrational partition functions can be approximated by

$$Q_{vib} = e^{-u/2}$$

Substituting numerical values, we obtain (in M^{-1} sec⁻¹ at 400 °K)

$$B(T) = 7.12 \times 10^{10}\Gamma_t \exp\left[-\tfrac{1}{2}(u^{\ddagger}_{stretch} + 2u^{\ddagger}_{bend} - u_{H_2})\right]$$

In evaluating Θ, the same assumption about u gives

$$\coth\left(\frac{u}{2}\right) = 1$$

and

$$\theta_{\text{vib}} = \left(\frac{u}{2}\right) - 1$$

thus

$$\Theta = \frac{1}{2}(4-1) + \left(\frac{u^{\ddagger}_{\text{stretch}}}{2}\right) - 1 + 2\left[\left(\frac{u^{\ddagger}_{\text{bend}}}{2}\right) - 1\right] - \left[\left(\frac{u_{\text{H}_2}}{2}\right) - 1\right] + \theta_t$$

$$= \frac{1}{2}[-1 + u^{\ddagger}_{\text{stretch}} + 2u^{\ddagger}_{\text{bend}} - u_{\text{H}_2}] + \theta_t$$

The Arrhenius pre-exponential factor is

$$A = 7.12 \times 10^{10} \, e^{-1/2} \Gamma_t e^\theta t$$

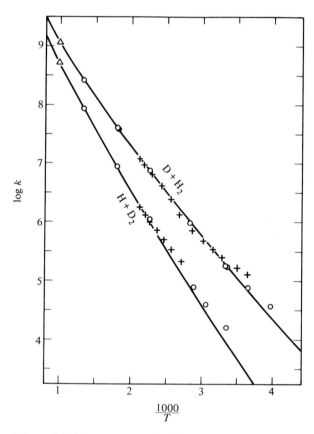

Figure 4.4. *Rate constants for the reactions* $D + H_2$ *and* $H + D_2$. *Solid line, calculated values; points, experimental values.* [*From Ref. 11b.*]

and the vibrational quantum corrections have exactly cancelled. Shavitt has carried out an analysis of the tunneling to be anticipated, and from this one finds $\Gamma_t = 3.11$ and $\theta_t = -1.84$. The final value for A is 2.14×10^{10} M^{-1} sec^{-1}.

By comparison, the hard-sphere collision model predicts an A value of 9.7×10^{10} if 0.934 Å is used for R. (This value is considerably smaller than the collision radius obtained from the transport properties of hydrogen gas.)

From the experimental data in this temperature region, one obtains a value of 0.9×10^{10}—about half the activated complex value.

The other three isotopic modifications of this reaction ($D + D_2$, $H + D_2$, and $D + H_2$) have also been considered by Shavitt. The theoretical potential energy surface predicts a barrier height of 10.994 kcal/mole; this value has been adjusted to 9.875 in order to bring about agreement between calculated and observed rate constants for $H + D_2$ and $D + H_2$ at temperatures near 1000 °K. A comparison of these same rate constants at lower temperatures is shown in Fig. 4.4, and the agreement is quite satisfactory.

Other calculations of these rate constants, based on semiempirical and empirical potential energy surfaces, are discussed by Johnston (12).

4.9. Example 3: Pre-Exponential Factors for Other Bimolecular Reactions

The A factors for several hydrogen-abstraction reactions have been calculated by Wilson and Johnston (13), using estimated bond lengths and force constants for the activated complex. The same approach had been previously applied (14) to a dozen reactions, most of them involving nitrogen oxides or oxyhalides. Similar estimates have been made for halogen atom reactions (15).

The general conclusion to be drawn from this work is that A factors for fairly complex reactions can be estimated, with about an order of magnitude as the largest difference between prediction and observation. A definite correlation of the A factor with molecular complexity is completely ignored by hard-sphere collision theory (unless it is taken into account by ad hoc adjustment of a steric "fudge factor").

4.10. Thermodynamic Formulation of the Rate Constant According to Activated Complex Theory

The equilibrium constant and the Gibbs free energy change ($\Delta G°$) for a chemical reaction $a\text{A} + b\text{B} \rightleftharpoons p\text{P} + r\text{R}$ are related to the partition functions for A, B, P, and R according to

$$-RT \ln K = \Delta G° = -RT \ln \left(\frac{Q_P^p Q_R^r}{Q_A^a Q_B^b} \right)$$

Therefore Eq. (4.19) for the rate constant can be written in the form

$$k = \left(\frac{\mathbf{k}T}{h} \right) K^{\ddagger} = \left(\frac{\mathbf{k}T}{h} \right) e^{-\Delta G^{\circ\ddagger}/RT} \tag{4.26}$$

where

$$\Delta G^{\circ\ddagger} = G° \text{ (activated complex)} - G° \text{ (reactants)}$$
$$= \Delta H^{\circ\ddagger} - T \Delta S^{\circ\ddagger}$$

Thus one speaks of the free energy of activation ($\Delta G^{\circ\ddagger}$), the enthalpy of activation ($\Delta H^{\circ\ddagger}$), or the entropy of activation ($\Delta S^{\circ\ddagger}$). Differences between other quasithermodynamic properties of the activated complex and reactants can also be defined (Section 6.9).

In the preceding equations, the entropy of activation depends on the choice of standard state (or the choice of concentration units), just as the rate constant itself does. The one exception is a unimolecular rate constant. The enthalpy and entropy of activation can be related to the usual Arrhenius parameters, again by using the experimental definition of activation energy:

$$E_a = \frac{-R \, d\ln k}{d(1/T)} = \frac{RT^2 \, d\ln k}{dT}$$

At constant pressure (appropriate for a condensed-phase reaction), thermodynamics gives the relation

$$\frac{d(G/T)}{dT_P} = \frac{-H}{T^2} \tag{4.27}$$

Hence

$$E_a = \Delta H^{\circ\ddagger} + RT \tag{4.28}$$

Equating the Arrhenius and thermodynamic forms of the rate constant, we have

$$A e^{-\Delta H^{\circ\ddagger}/RT} e^{-1} = \left(\frac{\mathbf{k}T}{h} \right) e^{\Delta S^{\circ\ddagger}/R} e^{-\Delta H^{\circ\ddagger}/RT}$$

whence

$$A = \left(\frac{\mathbf{k}T}{h}\right) \exp\left[\left(\frac{\Delta S^{\circ\ddagger}}{R}\right) + 1\right] \tag{4.29}$$

A reaction in the gas phase is customarily studied under conditions of constant volume, in which case one uses the thermodynamic relations

$$\frac{d(F/T)}{dT_V} = \frac{-E}{T^2}$$

$$F = G - PV$$

$$E = H - PV$$

If the perfect gas law can be assumed applicable to the reaction mixture,

$$\ln k = \text{constant} + \ln T - \frac{\Delta F^{\circ\ddagger} + \Delta(PV)^{\ddagger}}{RT}$$

$$= \text{constant} + \ln T - \frac{\Delta F^{\circ\ddagger}}{RT} + \Delta n^{\ddagger}$$

Then

$$\frac{RT^2\, d\ln k}{dT_V} = RT - T^2\frac{d(\Delta F^{\circ\ddagger}/T)}{dT_V}$$

and the activation energy is

$$E_a = RT + \Delta E^{\circ\ddagger}$$

$$= \Delta H^{\circ\ddagger} + RT\Sigma n_R^{\ddagger} \tag{4.30}$$

Here $\Sigma n_R^{\ddagger}$ is the molecularity of the reaction. The pre-exponential factor is

$$A = \left(\frac{\mathbf{k}T}{h}\right) \exp\left[\left(\frac{\Delta S^{\circ\ddagger}}{R}\right) + \Sigma n_R^{\ddagger}\right] \tag{4.31}$$

The entropy of activation can be interpreted in the usual way as an indication of the difference in disorder or randomness of the activated complex and the reactants.

4.11. The Effect of Isotopic Substitution on Rate Constants

Consider an equilibrium that exchanges isotopes:

$$AX + BX' \rightleftharpoons AX' + BX$$

where X and X' are different isotopes of the same element (H and D, C^{12} and C^{13}, etc.). The equilibrium constant in terms of partition functions is

$$K = \left(\frac{Q_{\mathrm{AX'}}}{Q_{\mathrm{AX}}}\right)\left(\frac{Q_{\mathrm{BX}}}{Q_{\mathrm{BX'}}}\right)$$ (4.32)

The ratio of *classical* partition functions for two isotopic varieties of the same molecule is given by Eq. (C.28):

$$\frac{Q_{\mathrm{AX'}}}{Q_{\mathrm{AX}}} = \frac{h^{3r}\int_{-\infty}^{\infty}\cdots\int \exp\left(-\Sigma p_i'^2/2m_i'kT\right)dp_1'\cdots dp_{3r}'}{h^{3r}\int_{-\infty}^{\infty}\cdots\int \exp\left(-\Sigma p_i^2/2m_ikT\right)dp_1\cdots dp_{3r}}$$

$$\times \frac{\int_{-\infty}^{\infty}\cdots\int \exp\left[-V'(q_1'\cdots q_{3r}')\right]dq_1'\cdots dq_{3r}'}{\int_{-\infty}^{\infty}\cdots\int \exp\left[-V(q_1\cdots q_{3r})\right]dq_1\cdots dq_{3r}}$$ (4.33)

According to the Born-Oppenheimer principle, the motion of electrons and nuclei can be considered separately. Therefore the electronic distribution in the molecule (which determines V) is unaffected by isotopic substitution, and the second ratio in (4.33) is unity. The integral over the momenta is a product of $3r$ one-dimensional integrals which are easily evaluated. The result is that

$$\frac{Q_{\mathrm{AX'}}}{Q_{\mathrm{AX}}} = \Pi_i\left(\frac{m_i'}{m_i}\right)^{3/2}$$

where the m_i are atomic masses.

A similar expression applies to the ratio $Q_{\mathrm{BX'}}/Q_{\mathrm{BX}}$. Since the same atoms appear on both sides of the chemical equation, the result is that there is no isotope effect, in the classical approximation;* that is,

$$K_{\mathrm{classical}} = 1$$

In discussing partition functions earlier in the chapter, we pointed out that both translational and rotational partition functions have classical values at ordinary temperatures. Therefore, in order to obtain the quantum version of Eq. (4.32), it is only necessary to put in the quantum corrections for vibrational partition functions, given in Eq. (C.25). The resulting equilibrium constant is

$$K = \frac{\Gamma(\mathrm{AX'})/\Gamma(\mathrm{AX})}{\Gamma(\mathrm{BX'})/\Gamma(\mathrm{BX})}$$ (4.34)

where

$$\Gamma = \prod_{i=1}^{3r-6} u_i e^{-u_i/2}(1 - e^{-u_i})^{-1}$$

*This omits a purely statistical factor arising from possible changes of rotational symmetry number caused by isotopic substitution.

Activated complex theory enables one to apply this method to the equilibrium between reactants and activated complex:

$$AX + B \rightleftharpoons [AXB]^{\ddagger} \longrightarrow products$$

$$AX' + B \rightleftharpoons [AX'B]^{\ddagger} \longrightarrow products'$$

The ratio of rate constants is, from Eq. (4.26),

$$\frac{k}{k'} = \frac{(\mathbf{k}T/h)K^{\ddagger}}{(\mathbf{k}T/h)K'^{\ddagger}} \tag{4.35}$$

In this expression, $K^{\ddagger} = Q_{3r-7}^{\ddagger}/Q_A Q_B$, and there is a similar equation for $K'^{\ddagger}$.

If the activated complex were a normal molecule, we could apply Eq. (4.33) again to find a ratio of partition functions

$$\frac{Q'^{\ddagger}_{3r-6}}{Q^{\ddagger}_{3r-6}} = \Pi_i \left(\frac{m'_i}{m_i}\right)^{3/2}$$

This step is not possible, however, because in (4.35) one degree of freedom of the activated complex has already been cancelled out in the ratio $(\mathbf{k}T/h)/(\mathbf{k}T/h)$. Therefore we must use instead

$$\frac{Q'^{\ddagger}_{3r-7}}{Q^{\ddagger}_{3r-7}} = \left(\frac{m^{\ddagger}}{m'^{\ddagger}}\right)^{1/2} \Pi_i \left(\frac{m'_i}{m_i}\right)^{3/2}$$

where $m^{\ddagger}$ is the reduced mass corresponding to motion along the reaction coordinate. Again the atomic masses are the same for activated complex and reactants; however, the classical rate constant ratio is not unity but rather

$$\left(\frac{k}{k'}\right)_{classical} = \left(\frac{m'^{\ddagger}}{m^{\ddagger}}\right)^{1/2} = \frac{v^{\ddagger}}{v'^{\ddagger}} \tag{4.36}$$

In this equation, $v^{\ddagger}$ is the vibrational frequency for motion along the normal coordinate, which coincides with the reaction coordinate at the saddlepoint, and it is obtained from the vibrational secular equation. It is *not* the frequency of passage over the barrier.

The quantum form of (4.36) requires the inclusion of quantum corrections for vibrations of A and B, for $3r - 7$ vibrations of $AXB^{\ddagger}$, and for the reaction coordinate (if tunneling is important). Thus

$$\frac{k}{k'} = \frac{v^{\ddagger}}{v'^{\ddagger}} \frac{\Gamma_t}{\Gamma'_t} \prod_{i=1}^{3r-7} \frac{\Gamma_i(AXB)^{\ddagger}}{\Gamma_i(AX'B)^{\ddagger}} \prod_{i=1}^{3r-6} \frac{\Gamma_i(AX')}{\Gamma_i(AX)} \tag{4.37}$$

The second term depends on the profile of the potential surface along the reaction path; all other terms depend on the vibrational frequencies —and ultimately, on force constants—or reactants and activated complex. In order for a kinetic isotope effect to exist, force constants of the activated complex must differ from those of the reactant, and these force constants must be related to motion of the atom X or X'.

If reactant force constants or vibrational frequencies are known from spectroscopy, $\Gamma(AX')/\Gamma(AX)$ can be calculated. Measurements of (k/k') then give information about the vibrational properties of the activated complex, which can be compared with properties predicted from various models of the potential energy surface.

An example is provided by the reaction $H_2 + Cl$, where rate constant ratios have been measured for the isotopic species HD, D_2, HT, T_2, and DT (16). The results are compared in Fig. 4.5 with predictions based on the BEBO method of calculating activated complex properties. The correct order of the various isotope effects is predicted, and the discrepancy between calculated and observed rate constant ratios is

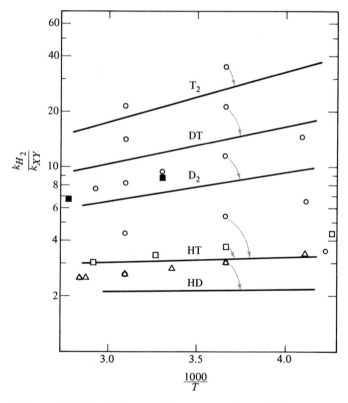

Figure 4.5. *Kinetic isotope effects in reactions of chlorine atoms with isotopic hydrogen molecules, XY. The ordinate is the ratio of rate constants for H_2 and species XY. Solid lines, values calculated from force constants obtained by the BEBO method (Section 2.7). Points, experimental values. [From Ref. 18a.]*

not large. This type agreement is one source of confirmation for the validity of the activated complex theory.

Kinetic isotope effects are also used in a more qualitative way to investigate reaction mechanisms. An early example is provided by the work of Melander (17) on electrophilic aromatic substitutions. Two possible mechanisms and the corresponding rate expressions are

I. $ArH + X^+ \longrightarrow ArX + H^+$

$\mathscr{R} = k(Ar)(X^+)$

and

II. $ArH + X^+ \underset{k_{-1}}{\overset{k_1}{\rightleftharpoons}} ArHX^+$

$ArHX^+ \overset{k_2}{\longrightarrow} ArX + H^+$

$\mathscr{R} = \left[\dfrac{k_1 k_2}{(k_{-1} + k_2)}\right](Ar)(X^+)$

These two mechanisms obviously cannot be distinguished by measuring the rate as a function of reactant concentrations.

Melander measured relative rates of H and T substitution by NO_2^+ in a series of aromatic molecules and found that this ratio did not differ significantly from unity. This observation rules out mechanism I: the breaking of a C—H bond should lead to a large difference between the force constants for this bond in ArH and in the activated complex. For the same reason, step 2 of mechanism II must not be rate-determining.

A number of reviews on kinetic isotope effects have been published (18).

References

1. W. J. Moore, *Physical Chemistry*, 3rd ed., Prentice-Hall, Englewood Cliffs, N.J., 1962, Chapter 15.
2. H. S. Johnston and P. Goldfinger, *J. Chem. Phys.* **37**, 700 (1962).
3. D. L. Bunker, *Theory of Elementary Gas Reaction Rates*, Pergamon Press, Oxford, 1966, Chapter 4.
4. L. Friedman, *Ann. Revs. Phys. Chem.* **19**, 273 (1968); J. H. Futrell and T. O. Tiernan, *Science* **162**, 415 (1968).
5. B. H. Mahan and J. C. Person, *J. Chem. Phys.* **40**, 392, 2851 (1964); T. S. Carlton and B. H. Mahan, *ibid.* **40**, 3683 (1964).
6. H. S. Johnston, *Gas-Phase Reaction Rate Theory*, Ronald Press, New York, 1966, Chapter 8.

7. R. A. Marcus, *Disc. Faraday Soc.* **44**, 7(1967); *J. Chem. Phys.* **46**, 959 (1967).
8. Reference 6, Chapter 2; E. F. Caldin, *Chem. Revs.* **69**, 135 (1969); H. S. Johnston and D. Rapp, *J. Am. Chem. Soc.* **83**, 1 (1961); H. S. Johnston and J. Heicklen, *J. Phys. Chem.* **66**, 532 (1962).
9. E. W. Schlag and G. L. Haller, *J. Chem. Phys.* **42**, 584 (1965); D. M. Bishop and K. J. Laidler, *ibid.* **42**, 1688 (1965).
10. Reference 6, Chapter 12.
11. I. Shavitt, R. M. Stevens, F. L. Minn, and M. Karplus, *J. Chem. Phys.* **48**, 2700 (1968); I. Shavitt, *ibid.* **49**, 4048 (1968).
12. Reference 6, Chapter 10.
13. D. J. Wilson and H. S. Johnston, *J. Am. Chem. Soc.* **79**, 29 (1957).
14. D. R. Herschbach, H. S. Johnston, K. S. Pitzer, and R. E. Powell, *J. Chem. Phys.*, **25**, 736 (1956).
15. G. C. Fettis, J. H. Knox, and A. F. Trotman-Dickenson, *J. Chem. Soc.* **1960**, 1064, 4177.
16. A. Persky and F. S. Klein, *J. Chem. Phys.* **44**, 3617 (1966).
17. L. Melander, *Arkiv Kemi* **2**, 211 (1950).
18. For example, R. E. Weston, Jr., *Science* **158**, 332 (1967); M. J. Goldstein, *ibid.* **154**, 1616 (1966); J. Bigeleisen, *ibid.* **147**, 463 (1965).

Problems

4.1. Derive the expression for the vibrational partition function, Eq. (C. 23), from (C. 19) and the following:

$$g_v = 1$$

$$E_v = (v + \tfrac{1}{2})h\nu \qquad \sum_{v=0}^{\infty} a^v = (1 - a)^{-1} \qquad \text{for } a < 1$$

[The index j of Eq. (C. 19) is equivalent to $v + 1$ in the preceding equations.]

4.2. Rate constants have been measured for the following reactions at 320 °K in the gas phase [J. M. Cambell and B. A. Thrush, *Trans. Faraday Soc.* **64**, 1265 (1968)]:

$$N + OH \longrightarrow NO + H \qquad k = 4.1 \times 10^{10} M^{-1} \sec^{-1}$$

$$O + OH \longrightarrow O_2 + H \qquad k = 3.0 \times 10^{10} M^{-1} \sec^{-1}$$

Compare the observed rate constants with those for hard-sphere collisions, assuming van der Waal's radii of 1.5 Å for N, 1.4 Å for O, and 1.5 Å for OH.

4.3. (*a*) Express the rate constant for a reaction in solution in terms of entropy of activation and activation energy.

(*b*) The rate constant for the reaction

$$Fe^{2+} + Fe(phen)_3^{3+} \longrightarrow Fe^{3+} + Fe(phen)_3^{2+}$$

(where phen = 1,10–phenanthroline) is $3.70 \times 10^4\ M^{-1} \sec^{-1}$ at 25°C. in 0.5 M HClO$_4$. The activation energy, found from a plot of log k vs. T^{-1}, is 0.8 kcal/mole. What are the entropy and enthalpy of activation? [N. Sutin and B. M. Gordon, *J. Am. Chem. Soc.* **83**, 70 (1961.)]

4.4. Molecules in a gas with a Maxwell-Boltzmann velocity distribution collide with the *inert* surface of the container at a rate $\mathcal{R}$ (molecules sec^{-1} cm^{-2}). Use the activated complex method to find an expression for $\mathcal{R}$.

Hint: Define the activated complex as a molecule at a distance of one mean free path from the surface but with internal energy distribution identical with that of a molecule in the gas phase.

4.5. The recombination of radicals can be treated as the collision of particles subject to the attractive part of a Lennard-Jones potential, $V(r) = -4\epsilon(\sigma_{LJ}/r)^6$.

(a) Derive the expression for the rate constant, assuming radicals are at thermal equilibrium.

Hint: The definite integral

$$\int_0^\infty x^m e^{-ax^2}\, dx = \frac{\Gamma[\frac{1}{2}(m+1)]}{2a^{(m+1)/2}}$$

(b) Use this result to calculate the rate constant for methyl radical recombination at 400 °K from the parameters of Problem 8, Chapter 3.

Hint: $\Gamma(5/3) = 0.90274$.

4.6. The rate derived in Problem 4.4 is also the rate of effusion—that is, the rate at which gas molecules escape through a hole (small compared to the distance between molecular collisions) in a thin diaphragm.

(a) This process is applied to gaseous UF_6 in order to separate the fissionable isotope U^{235} from U^{238}. What is the concentration of $U^{235}F_6$ relative to that of $U^{238}F_6$ in the effluent gas?

(b) Show that the mean kinetic energy of the molecules *that effuse* is $2kT$ per molecule, compared with $3kT/2$ for a molecule within the container.

Hint: If dA is the area of the hole and θ is the angle between the molecular velocity vector and the normal to the hole, then molecules within a volume $c \cos \theta\, dt\, dA$ will reach the hole in a time interval dt. Combine this with $P(c)\, dc$ to average the kinetic energy.

4.7. For sufficiently small values of $u^\ddagger(= hv^\ddagger/kT)$, it can be shown that the tunneling correction becomes

$$\Gamma_t = 1 + \left(\frac{u^{\ddagger 2}}{24}\right)$$

Show that, at high temperatures, the kinetic isotope effect has the form

$$\ln\left(\frac{k}{k'}\right) = A + BT^{-2}$$

4.8. Extremely precise rate measurements (for reactions in solution) indicate that $\Delta H^{\circ\ddagger}$ and $\Delta S^{\circ\ddagger}$ are *not* independent of temperature. This factor leads to the concept of the heat capacity of activation, $\Delta c_p^{\circ\ddagger}$. At constant pressure this quantity is defined by

$$\Delta c_p^{\circ\ddagger} = c_p^\circ \text{ (activated complex)} - c_p^\circ \text{ (reactants)}.$$

Over a reasonable temperature range $\Delta c_p^{\circ\ddagger}$ can be expressed as

$$\Delta c_p^{\circ\ddagger} = A + BT$$

Using the experimental definition of the activation energy and Eq. (4.27), find E_a as a function of temperature.

4.9. The potential energy surface for the reaction $H_2 + Br \longrightarrow HBr + H$ has been obtained with the LEPS method [R. B. Timmons and R. E. Weston, Jr., *J. Chem. Phys.* **41**, 1654 (1964)]. The activated complex is linear, with the following properties:

Vibrational frequencies, 2472, 217 (doubly degenerate), $178i$ cm^{-1}

$R(H\text{---}H) = 1.62$ Å, $R(H\text{---}Br) = 1.42$ Å

$V_a = 18.1$ kcal/mole

$g_{el} = 2$

Necessary reactant properties are $R(H\text{---}H) = 0.742$ Å, $v = 4395$ cm^{-1}, $g_{el} = 1$. For Br, $g_{el} = 4$.

(a) Use the activated complex formulation to calculate A and E_a at 500 °K. Assume that tunneling can be neglected.

(b) Vibrational frequencies for the analogous D–D–Br complex are 1760, 154(2), and $127i$ cm^{-1}. The vibrational frequency of D_2 is 3108 cm^{-1}. Calculate relative rate constants for $Br + H_2$ and D_2 at 500 °K.

5

Unimolecular

Reactions

IN THIS CHAPTER we shall discuss the class of elementary reactions that consist of the decomposition or isomerization of an individual molecule. At high pressures, the empirical rate expression is first order with respect to the reacting species; this concentration dependence gradually changes to second order as the pressure is decreased. For this reason, such reactions are sometimes categorized as "quasiunimolecular."

5.1. Spontaneous Decomposition of Energized Molecules

We have already commented on the temperature dependence of rate constants and have explained this factor in terms of an energy barrier separating reactants from products. For a hypothetical unimolecular

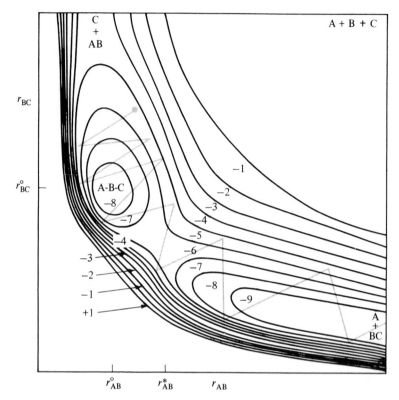

Figure 5.1. *Potential energy contours for the unimolecular decomposition*
ABC (linear) ⟶ *A + BC. Equilibrium bond lengths of ABC are r°_{AB} and*
r°_{BC}, while r^{*}_{AB} represents the critical length of this bond. The solid line
represents vibrations leading to decomposition.

decomposition

$$ABC \longrightarrow A + BC$$

this barrier is depicted qualitatively in Fig. 5.1, which shows the potential
energy surface when the three atoms are confined to a linear configuration. The region of stability of ABC is indicated by the basin with a
minimum at r°_{AB} and r°_{BC}. If an ABC molecule somehow acquires enough
energy, the AB bond can stretch to the point where there is no restoring
force (designated as r^{*}_{AB}); here the bond breaks and the products A and
BC are formed. A similar surface can be drawn for an isomerization
process.

5.2. Formation of Energized Molecules

Now the question arises: How does the reactant molecule obtain sufficient energy to undergo the process just described? In view of the applicability of the kinetic theory of gases to bimolecular processes, it is reasonable to assume that the necessary energy is obtained when the reactant collides with another molecule with a high relative kinetic energy. The question of how this bimolecular process can lead to a first-order reaction still remains, however. The solution (1) was proposed by F. A. Lindemann (later Lord Cherwell) in 1922, and the important concept is the existence of a time lag between the moment of the bimolecular collision and the time at which the energized molecule decomposes. During this time, the molecule can undergo another collision and lose its energy.

The scheme derived from Lindemann's proposal can be written as follows:

$$A + M \xrightarrow{k_a} A^* + M \quad \text{activation}$$

$$A^* + M \xrightarrow{k_d} A + M \quad \text{deactivation} \tag{5.1}$$

$$A^* \xrightarrow{k_E} B \quad \text{reaction of energized molecule}$$

Here A is the reactant, A^* is an "energized" reactant molecule—that is, one with enough energy to react—and M is either another molecule of reactant or a molecule of some added inert gas. The rate is given by

$$\frac{d(B)}{dt} = k_E(A^*)$$

The steady-state method can be used to solve for the steady-state concentration of A^* in the usual way:

$$\frac{d(A^*)}{dt} = 0 = k_a(A)(M) - k_d(A^*)(M) - k_E(A^*)$$

whence

$$(A^*) = \frac{k_a(A)(M)}{[k_d(M) + k_E]}$$

Hence the rate is simply

$$\frac{d(B)}{dt} = \frac{k_a k_E(A)(M)}{k_d(M) + k_E} = k_{uni}(A) \tag{5.2}$$

where k_{uni} is the experimental first-order rate constant.

It is apparent that k_{uni} will depend on the concentration of M or the total pressure; however, there are two limiting cases that lead to particularly simple forms of k_{uni}. At high pressures, if $k_d(M) \gg k_E$, we

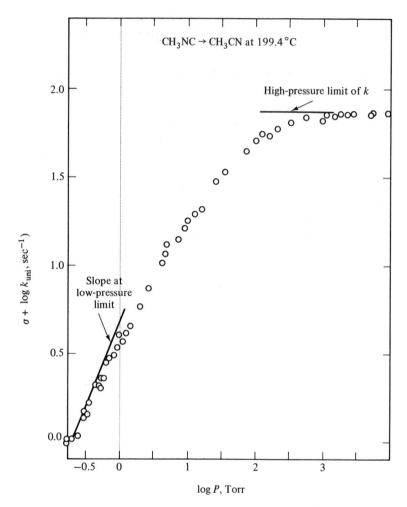

Figure 5.2. *Pressure dependence of the observed first-order rate constant* k_{uni}, *for the isomerization* $CH_3NC \longrightarrow CH_3CN$ *at 199.4 °C. [Data from Ref. 2.]*

find that the experimental rate constant is

$$k_\infty \equiv (k_{uni})_{p=\infty} = \frac{k_a}{k_d} k_E \tag{5.3}$$

The empirical quantity k_∞ is independent of pressure, so that the reaction is indeed first order. In this case, the rate of reaction of energized molecules is so low compared with the rate of energy removal in the second step that an equilibrium fraction of $(A^*)/(A) = k_a/k_d$ is maintained. Thus the empirical rate constant is merely the product of the rate constant for spontaneous decomposition and the fraction of molecules that are energized. At low pressures, the opposite inequality applies—that is, $k_d(M) \ll k_E$—and we find that the experimental rate constant is

$$k_0 \equiv (k_{uni})_{p=0} = k_a(M) \tag{5.4}$$

In this limiting pressure region, the rate–determining step has become the production of energized molecules by collision. We expect to find that the quantity k_{uni} is linearly dependent on the total pressure. The products of the reaction are generally about as effective as the reactant in transferring energy during a collision, so that even in the absence of added foreign gas, (M) will be nearly constant and the rate expression is first order during a given experiment.

These important predictions of this oversimplified form of unimolecular reaction theory are indeed confirmed experimentally. Figure 5.2 shows the pressure dependence of k_{uni} for the isomerization of methyl isocyanide to methyl cyanide (2), plotted in logarithmic form. At high pressures, the line asymptotically approaches $\log k_\infty$; and at low pressures, the line has a slope of unity, corresponding to the region in which Eq. (5.4) is applicable.

5.3. *Variation of Rate Constants with Energy*

The preceding discussion is oversimplified because it neglects the variation of the rate constants k_a, k_d, and k_E with energy. But on the basis of the preceding discussion of potential energy surfaces, it is clear that k_E must increase with increasing energy. Indeed, it was given the subscript E in anticipation of this fact. The earliest attempt to take· this factor into account was that of Hinshelwood (3), who assumed that

$$k_E = 0 \qquad E < E_c$$

$$k_E = \text{constant} \qquad E \geq E_c$$

where E_c may be called the "critical" energy. The result leads to the same simple expression previously found as Eq. (5.2) and does give the correct pressure dependence at the high- and low-pressure limits. However, between these limits, the effect of pressure is not predicted correctly. Thus rearrangement of Eq. (5.2) leads to

$$\frac{1}{k_{uni}} = \frac{k_d}{k_a} k_E + \frac{1}{k_a(M)}$$

so that a plot of $1/k_{uni}$ against $1/P$ should give a straight line; this behavior is actually found only in the immediate vicinity of the high-pressure limit (Fig. 5.3).

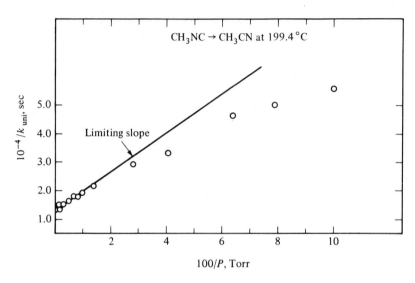

Figure 5.3. *Behavior of $1/k_{\text{uni}}$ for the methyl isocyanide isomerization near the high-pressure limit.*

Before considering specific models of energized molecules, let us obtain the energy dependence of k_{uni} in a general form. We retain our notation for k_E, with the understanding that this is the rate constant at a specific energy E. Alternatively, if energy is considered to be quantized rather than continuous, the symbol k_j will be used to indicate the rate constant for the jth energy level of the molecule. The development then follows that leading to Eqs. (5.2) to (5.4). Similarly, k_{aj} and k_{dj} are rate constants for activation and deactivation of A molecules in the jth energy level. At concentrations high enough so that the loss of energized molecules by dissociation is negligible (i.e., the high-pressure limit) the fraction of energized molecules in a given energy level j ("j molecules") is just

$$\frac{(A_j^*)}{(A)} = \frac{k_{aj}}{k_{dj}} = P_j \tag{5.5}$$

The rate of unimolecular reaction is then given by the sum over all energy levels of the rates for individual levels; each individual rate depends on the population of the level and the rate constant appropriate to it. Thus

$$\text{Total rate} = k_\infty(A) = \sum_{j=0}^{\infty} k_j(A_j^*) = \sum_j k_j P_j(A) \tag{5.6}$$

so that

$$k_\infty = \sum_j k_j P_j \qquad (5.7)$$

An equivalent form, suitable if the energy is considered to be a continuous variable, is given by

$$k_\infty = \int_0^\infty k_E P(E)\, dE \qquad (5.8)$$

where $P(E)\, dE$ is the fraction of molecules with energy in the range E to $E + dE$.

At this point it is useful to consider the processes of activation (or energization) and deactivation by collision in more detail. The assumption almost universally made (for which some justification will be given later in this chapter) is that of "strong collisions." Collisions of energized molecules with each other can be neglected, due to their low concentration; this assumption is implied in the mechanism of Eq. (5.1). Each collision of an energized molecule with a ground-state molecule (or a molecule of foreign gas) is assumed to result in a large enough transfer of energy so that the energy content of *neither* molecule is sufficient for reaction to occur. This assumption makes it possible to replace the energy-dependent constant k_{dj} by a rate constant for bimolecular collisions, k_{coll}, which is derived from the usual kinetic theory expressions (page 87). Uncertainties in collisional cross sections determined by various types of measurement (viscosity, etc.) will not be very important. One can also use the *frequency* of collisional deactivation ω, defined as

$$\omega \equiv Z_{AM}(M) \quad \text{or} \quad Z_{AA}(A)$$

Then the rate at which j molecules are de-energized is

$$\omega P'_j(A)$$

where the prime indicates that this is a steady-state, and not an equilibrium, energy distribution in the most general case.

Finally, we consider the rate of energization or the rate of formation of j molecules from ground-state molecules. Returning temporarily to the high-pressure limit, we see that the reverse process of de-energization would take place at a rate

$$\omega P_j(A)$$

By the principle of detailed balancing, molecules with less than the critical energy will be energized by collision to the level j at this same rate. Now, because of decomposition, the actual system has a smaller population of j molecules than does the high-pressure, equilibrium system. But the distribution of molecules in states below the critical energy will be almost the same in both cases, since the molecules in lower states have no way of "anticipating" whether they will ultimately react or not. (It is assumed in the foregoing discussion that the rate of production of j molecules from ground-state molecules is much greater than the rate of production from molecules in still higher states. This is another aspect of the strong collision hypothesis.) Thus the rate of energization by collision is given by the preceding expression, even when steady-state, rather than equilibrium, conditions prevail.

The foregoing rates of appearance and disappearance of j molecules are combined to give

$$k_j P'_j(A) + \omega P'_j(A) = \omega P_j(A)$$

Hence the steady-state distribution function is

$$P'_j = \frac{\omega P_j}{\omega + k_j} \tag{5.9}$$

The unimolecular rate constant, obtained as before by summation over all energy levels, is

$$k_{\text{uni}} = \sum_j k_j P'_j = \frac{\sum_j \omega k_j P_j}{(\omega + k_j)} \tag{5.10}$$

The corresponding differential form is

$$k_{\text{uni}} = \frac{\int_{E_c}^{\infty} \omega k_E P(E)\, dE}{(\omega + k_E)} \tag{5.11}$$

These expressions, valid at any pressure, reduce to the previously derived forms of k_∞ as ω/k_E becomes very large. The evaluation of either Eq. (5.10) or (5.11) depends on the specific forms of the rate constant k_j or k_E and the distribution function P_j or $P(E)$.

5.4. The Rice-Ramsperger and Kassel Formulations of k_{uni}

Explicit formulations of $P(E)$ and k_E for a classical model, and P_j and k_j for a quantized model, were developed independently by Rice and Ramsperger and by Kassel (4,5). Since the quantum version of Kassel is more realistic and can be shown to lead to the classical version under suitable conditions, we shall discuss it in some detail.

The model assumed for the reactant molecule is a system of s identical harmonic oscillators, each of frequency v. These oscillators are said to be "loosely coupled": they interact enough so that energy can flow from one oscillator to another, but not strongly enough to perturb each other's energy levels. Thus the reactant gas is a double statistical assembly: at the *inter*molecular level, molecules interchange energy at every collision; whereas at the *intra*molecular level, energy is interchanged randomly among the oscillators, between the times of molecular collisions. The total energy of the energized molecule is determined by specification of j, the total number of quanta, so that

$$E_j = jhv \tag{5.12}$$

Then the fraction of molecules in the jth level is

$$P_j = \frac{N_j}{N} = \frac{g_j \exp\left(-jhv/kT\right)}{\sum\limits_{i=0}^{\infty} g_i \exp\left(-ihv/kT\right)} \tag{5.13}$$

The statistical weight g_j is determined by the degeneracy of the jth energy level; it is the number of different ways the j quanta can be distributed among the s oscillators and is given by*

$$g_j = \frac{(j+s-1)!}{j!(s-1)!} \tag{5.14}$$

The summation in the denominator of Eq. (5.13) is the appropriate partition function Q, and it is given by

$$
\begin{aligned}
Q &= \sum_{i=0}^{\infty} \frac{(i+s-1)!}{i!(s-1)!} e^{-iu} \\
&= \frac{(s-1)!}{(s-1)!} + \frac{s!}{(s-1)!} e^{-u} + \frac{(s+1)!e^{-2u}}{2!(s-1)!} + \cdots \\
&= 1 + se^{-u} + \frac{s(s+1)e^{-2u}}{2!} + \cdots
\end{aligned}
\tag{5.15}
$$

In the foregoing expressions, $u = hv/kT$. The final form of (5.15) may be recognized as the binomial theorem expansion of $(1 - e^{-u})^{-s}$, so that

$$P_j = \frac{(j+s-1)!(1-e^{-u})^s e^{-ju}}{j!(s-1)!} \tag{5.16}$$

According to the assumptions of this same model, the rate constant k_j is proportional to the probability that m of the total j quanta contained in the s oscillators are in a particular fixed distribution. A reasonable physical model, although more restrictive than necessary mathematically, would be for the m quanta to be in one oscillator corresponding

---

*The traditional analog is as follows: In how many ways can one distribute j balls among s boxes? The s boxes can be represented by $s-1$ partitions. If the j balls and $s-1$ partitions are placed on a line, there are $(j+s-1)!$ possible permutations of these objects. But since all j balls are alike and $s-1$ boxes are alike, we must divide by $j!$ and $(s-1)!$ to account for identical configurations. Check (5.14) with small values of j and s to convince yourself that the procedure is correct.

to the bond that is breaking. The quantity mhv is the critical energy, E_c, in this model. We are able to use the same combinatorial procedure as that used for g_j if we note that with m of the j quanta fixed, there are $j - m$ quanta to be distributed among the remaining $s - 1$ oscillators. The number of ways of doing so is simply

$$\frac{(j - m + s - 1)!}{(j - m)!(s - 1)!} \tag{5.17}$$

However, this factor is too large because it also includes the number of ways in which the total j quanta can be partitioned among s oscillators. Dividing by the factor g_j to correct for (5.17) being too large, we have (for $j \geq m$)

$$
\begin{aligned}
k_j &= \frac{D(j - m + s - 1)!}{(j - m)!(s - 1)!} \frac{j!(s - 1)!}{(j + s - 1)!} \\
&= D\frac{j!(j - m + s - 1)!}{(j - m)!(j + s - 1)!}
\end{aligned}
\tag{5.18}
$$

In this expression D is a proportionality factor with dimensions of sec^{-1}. Its physical significance is that of a rate constant for the exchange of energy among the oscillators.

The classical expressions equivalent to Eqs. (5.16) and (5.18) can be obtained by making use of the correspondence between classical mechanics and quantum mechanics that is valid when u is small or when j becomes so large that it is essentially continuous. In the present case, j is therefore much larger than either m or s at this classical limit. We note then that the expression for g_j simply becomes*

$$
\lim_{j \to \infty} g_j \equiv G(E)
$$

$$
= \lim_{j \to \infty} \left[\frac{(j + s - 1)!}{j!(s - 1)!} \right] = \frac{j^{s-1}}{(s - 1)!} \tag{5.19}
$$

When u is small, the term in P_j that is the harmonic oscillator partition function $(1 - e^{-u})$ can be replaced by its limiting value hv/kT. Moreover, we recall that $E = jhv$, and since j is essentially continuous,

$$
\frac{dE}{dj} = \frac{d(jhv)}{dj} = hv
$$

---

*Consider the expression

$$\frac{(x + y)!}{x!}$$

when $x \gg y$. Then this is

$$\frac{x!(x + 1)(x + 2) \cdots (x + y)}{x!}$$

But if y is much smaller than x, so is $y - 1, y - 2, \ldots, 1$; thus the y terms remaining in the fraction are approximately equal to x^y.

Combining these factors, we find the classical form of Eq. (5.16)

$$P(E)\, dE = \lim_{j \to \infty} P_j\, dj = \frac{(h\nu/kT)^s e^{-E/RT}(E/h\nu)^{s-1}(dE/h\nu)}{(s-1)!}$$

$$= \frac{E^{s-1} e^{-E/RT}\, dE}{(RT)^s (s-1)!} \tag{5.20}$$

In terms of a dimensionless quantity $y = E/RT$, this expression may be written

$$P(y)\, dy = \frac{y^{s-1} e^{-y}\, dy}{(s-1)!} \tag{5.21}$$

The result obtained in this way is identical with that obtained from the more rigorous (but much more complicated) derivation originally given in terms of classical phase integrals.

The quantum and classical forms of the distribution function are shown in Fig. 5.4 as a function of y for the case where $s = 10$. Although s is the only parameter required in the classical model, in order to evaluate the quantum P_j one must also have a value of $u = h\nu/kT$, and this requirement is a disadvantage of the quantum formulation. The data illustrated are for the case where $u = 1$, so that $j = y$. This is not an

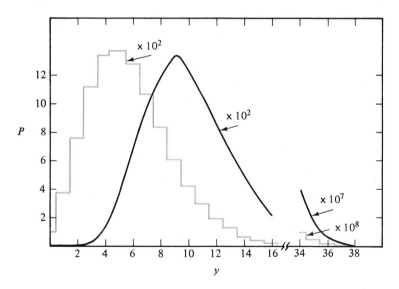

Figure 5.4. *Molecular distribution functions plotted against $y = E/RT$. Smooth curve, $P(E)$ from the classical RRK expression; step function, P_j from the Kassel quantum form. Parameters are $s = 10$, $u = 1$.*

unreasonable value to choose: it corresponds to an oscillator frequency of about 2×10^{13} sec^{-1} (700 cm^{-1}) at a temperature of 1000 °K. It is evident that these conditions are a poor approximation to the classical limit, and it will only be for unreasonably small values of u that the two methods converge. The problem of evaluating P_j and $P(E)$ will be discussed again later in the chapter.

Nevertheless, the classical approximation is important in giving one a physical understanding of the unimolecular process. The function $P(y)$ qualitatively resembles a Maxwellian velocity distribution function. It can be shown that the average energy of all reactant molecules is just sRT, and that the distribution function peaks at a value of y equal to $s - 1$.

The analogous classical expression for the decomposition rate constant is easily obtained by the same approach. We again allow j to become very large compared with s or m, so that $j - m$ is much larger than s. In this case,

$$k_E = \lim_{j \to \infty} k_j = \lim_{j \to \infty} \left[\frac{Dj!(j - m + s - 1)!}{(j - m)!(j + s - 1)!} \right]$$

$$= \frac{D(j - m)^{s-1}}{j^{s-1}} = D \left(\frac{E - E_c}{E} \right)^{s-1} \tag{5.22}$$

In the last step, we have made use of the definitions

$$E = jh\nu \quad \text{and} \quad E_c = mh\nu$$

Again, this same result can be obtained by an approach that is completely classical from the beginning.

To illustrate the energy dependence of k_E (Fig. 5.5), we have chosen a value of 35 for $y_c = (E_c/RT)$; this value is about what is required to give a first-order rate constant within the range of experimental determination. Again a value of $s = 10$ was used for this illustration. It is apparent that the classical RRK model predicts that k_E rises very steeply from zero at the critical energy E_c and only begins to approach the limiting value of D at energies far beyond the critical energy. Of course, the behavior of k_E depends on the number of oscillators; for smaller values of s the energy dependence will be less dramatic, as is illustrated by Table 5.1. The way in which k_E at a particular value of $E = E_c + sRT$ depends on the parameters s and E_c/RT is shown in Table 5.2. For a given critical energy, the value of k_E/D drops off rapidly as s increases, so that one would expect the average lifetime of an energized molecule to be longer the greater the number of oscillators it contained. Indeed, a correlation between lifetime and the number of atoms in the molecule is found experimentally. The physical explanation is that there are more ways of distributing the energy, and hence less chance for the critical

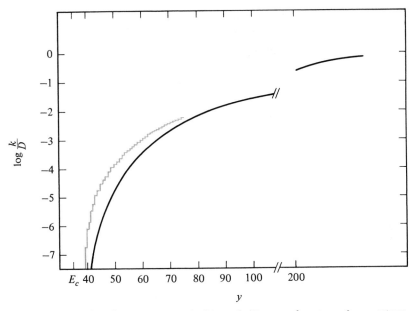

Figure 5.5. *Reduced rate constant k_E/D or k_j/D as a function of $y = E/RT$. Smooth curve, k_E/D from the classical RRK expression; jagged line, k_j/D from the Kassel quantum form. Parameters are $s = 10$, $u = 1$, and $E_c/RT = 35$.*

energy to become localized, if s is large. As one would expect, for a given value of s, the rate constant decreases as the critical energy is increased.

Application of the quantum version of this model again requires that another parameter be fixed, and we have assumed that $u = 1$ as in the preceding discussion of P_j. This version predicts a finite (but extremely small) value of k_j at the critical energy (or more accurately, when $j = m$); in fact, the rate constant is merely Dg_m^{-1}. The calculated values of k_j are consistently larger than values of k_E from Eq. (5.22), but the two approaches begin to converge at very high energies. Once again, the

Table 5.1. *Values of E/E_c at which $k_E = D/2$*

$s-1$	E/E_c
2	3.4
4	6.3
6	9.2
10	15
20	29
30	44

Table. 5.2. *The Quantity k_E/D at $E = E_c + sRT$ as a Function of s and y_c $(= E_c/RT)$*

s	$y_c =$	20	30	40
2		$9.2(-2)^a$	$6.3(-2)$	$4.8(-2)$
4		$4.6(-3)$	$1.6(-3)$	$7.5(-4)$
6		$6.5(-4)$	$1.3(-4)$	$3.8(-5)$
8		$1.6(-4)$	$2.1(-4)$	$3.6(-6)$
10		$5.1(-5)$	$3.8(-6)$	$5.1(-7)$
15		$7.0(-6)$	$2.1(-7)$	$1.3(-8)$
21		$1.5(-6)$	$2.0(-8)$	$5.5(-10)$

^{a}Numbers in parentheses are powers of ten by which coefficients of k_E/D are to be multiplied.

classical approximation is a poor one at values of u as large as unity but provides a qualitative understanding of the energy dependence.

5.5. Rate Constants at the High-Pressure Limit

It is instructive to consider the form assumed by the RRK unimolecular rate constant at pressures high enough so that all values of k_E or k_j are much smaller than the collision frequency ω. Then the steady-state distribution function P'_j is identical with the equilibrium value P_j. Equation (5.7) is evaluated to give

$$k_\infty = D \sum_{j=m}^{\infty} \frac{(j - m + s - 1)!}{(j - m)!(s - 1)!}(1 - e^{-u})^s e^{-ju}$$

$$= D(1 - e^{-u})^s e^{-mu} \sum_{r=0}^{\infty} \frac{(r + s - 1)!}{r!(s - 1)!}e^{-ru} \qquad (5.23)$$

where $r = j - m$. The summation has already been evaluated in Eq. (5.16); it is simply $(1 - e^{-u})^{-s}$, so that

$$k_\infty = De^{-mu} \qquad (5.24)$$

The classical expression (Eq. 5.8) becomes

$$k_\infty = \int_{E_c}^{\infty} k_E P(E) \, dE$$

$$= \int_{E_c}^{\infty} \frac{D[(E - E_c)/E]^{s-1}E^{s-1}e^{-E/RT} \, dE}{(RT)^s(s - 1)!} \qquad (5.25)$$

Equation (5.25) can be evaluated by using the variable $x = (E - E_c)/RT$, so that

$$k_\infty = \frac{De^{-E_c/RT}}{(s - 1)!} \int_0^{\infty} x^{s-1}e^x \, dx$$

This is a standard definite integral with the value $(s - 1)!$ Therefore the limiting value of the rate constant is simply

$$k_\infty = De^{-E_c/RT} \tag{5.26}$$

The rate constant in both Eqs. (5.24) and (5.26) are obviously of the Arrhenius form. According to this model, the experimental activation energy E_∞ is merely the critical energy E_c, while the pre-exponential factor gives the quantity D, which was originally introduced as an arbitrary proportionality constant.

It should also be pointed out that Eqs. (5.23) and (5.25) are expressions that average the rate constant over a distribution function characteristic of the equilibrium system. In the case of the classical form, the energy dependence of the function $k_E P(E)$ is shown in Fig. 5.6, and the area under this curve represents k_∞. Because k_E is rising sharply and $P(E)$ is dropping rapidly at energies above E_c, the function is strongly peaked around a maximum of $(s - 1)RT + E_c$, and one can show that the average energy of reacting molecules is just $E_c + sRT$. It is clear that the difference between the average energy of reacting molecules $\bar{\bar{E}}$ and the average energy of all molecules $\bar{E}$ is just E_c, the activation energy. This is an example of a general theorem due to Tolman (6):

$$E_a = \bar{\bar{E}} - \bar{E} \tag{5.27}$$

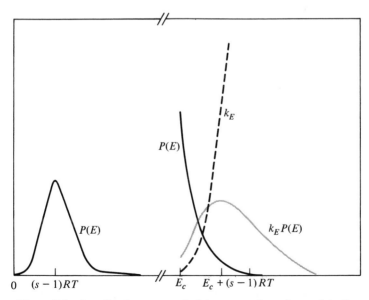

Figure 5.6. *A qualitative portrayal of the energy dependence of the function* $k_E P(E)$. *The ordinate is greatly magnified in the region above* E_c.

Table 5.3. *Arrhenius Parameters for Some Unimolecular Reactions*

Reactant	Products	$\log A_\infty$ (A in sec^{-1})	E_∞ (kcal/mole)	Reference
1. ISOMERIZATIONS				
CH_3NC	CH_3CN	13.6	38.4	a
cis–$CHD{=}CHD$	*trans*–$CHD{=}CHD$	12.5	61.3	b
cyclo–C_3H_6	CH_3–$CH{=}CH_2$	15.2	65.0	c
CH_3–cyclo–C_3H_6	$\begin{cases} CH_2{=}CHCH_2CH_3 \\ \textit{cis–} \text{and } \textit{trans–} \\ CH_3\text{–}CH{=}CHCH_3 \end{cases}$	15.4	65.0	d
2. DECOMPOSITIONS TO STABLE MOLECULES				
C_2H_5Cl	$C_2H_4 + HCl$	14.6	60.8	e
cyclo–C_4H_8	$2C_2H_4$	15.6	62.5	f
cyclo–C_4F_8	$2C_2F_4$	16.0	74.3	g
C_2H_5—O—$CH{=}CH_2$	$C_2H_4 + CH_3CHO$	11.4	43.8	h
3. DECOMPOSITIONS TO RADICALS				
N_2O_5	$NO_2 + NO_3$	14.8	21	i
C_2H_6	$2CH_3$	17.4	91.7	j
$CH_3N{=}NCH_3$	$2CH_3 + N_2$	17.2	55.5	k
$(CH_3)_3COOC(CH_3)_3$	$2(CH_3)_3CO$	15.6	37.4	l

[a]F. W. Schneider and B. S. Rabinovitch, *Journal of the American Chemical Society* **84**, 4215 (1962).

[b]B. S. Rabinovitch et al., Journal of Chemical Physics **20**, 1807 (1952); **23**, 315, 2439 (1955).

[c]H. O. Pritchard, R. G. Sowden, and A. F. Trotman-Dickenson, *Proceedings of the Royal Society (London)* **A217**, 563 (1953).

[d]J. P. Chesick, *J. Am. Chem. Soc.* **82**, 3277 (1960).

[e]K. E. Howlett, *Journal of the Chemical Society* (London) 3695, 4487 (**1952**).

[f]C. T. Genaux, F. Kern, and W. D. Walters, *J. Am. Chem. Soc.* **75**, 1966 (1953).

[g]B. Atkinson and A. B. Trenwith, *J. Chem. Phys.* **20**, 754 (1952); J. M. Butler, *J. Am. Chem. Soc.* **84**, 1393 (1962).

[h]A. T. Blades and G. W. Murphy, *J. Am. Chem. Soc.* **74**, 1039 (1952).

[i]R. L. Mills and H. S. Johnston, *J. Am. Chem. Soc.* **73**, 938 (1951); Johnston, *ibid.*, 4542.

[j]C. P. Quinn, *Proc. Royal Soc. (London)* **A275**, 190 (1963).

[k]W. Forst and O. K. Rice, *Canadian Journal of Chemistry* **41**, 562 (1963).

[l]L. Batt and S. W. Benson, *J. Chem. Phys.* **36**, 895 (1962).

The Arrhenius parameters A_∞ and E_∞ (or D and E_c, according to this model) are evaluated experimentally in the following way. At a given temperature, the unimolecular rate constant is measured at a number of pressures, and a plot like that of Fig. 5.3 is used to obtain k_∞ by extrapolation. (For a discussion of alternative extrapolation procedures, see Reference 7.) Then the usual Arrhenius plot is made of log k_∞ against T^{-1}, and the slope and intercept give E_∞ and A_∞, respectively. A few data, selected from the large number available, are presented in Table 5.3.

Activation energies for unimolecular reactions are generally fairly high*; in the case of bond-fission reactions, they are related to the enthalpies of breaking the bonds and provide a means for determining these quantities according to the following procedure.

Consider the unimolecular fission into radicals

$$A\!-\!X\!-\!Y\!-\!B \longrightarrow AX + YB$$

The enthalpy ΔH of this reaction is often called the "bond-dissociation energy of the X—Y bond."† The enthalpy change is related, in turn, to the activation energies of the forward and reverse reactions (E_{af} and E_{ar}) according to

$$\Delta H = \Delta E + RT = E_{af} - E_{ar} + RT$$

The term RT enters because of the change in the number of moles, but it is usually neglected in practice. We have seen previously that activation energies for radical recombinations are very small or zero, so that $\Delta H \cong E_{af}$, and the bond-dissociation energy is approximately equal to the activation energy for the unimolecular bond rupture. Further details of the experimental techniques, as well as tables of bond-dissociation energies obtained by this method, are given by Kerr (8).

5.6. Rate Constants at the Low-Pressure Limit

At the other pressure limit, when $\omega \ll k_E$ or k_j for all values of the energy, the unimolecular rate constant becomes

$$k_0 = \omega \sum_{j \geq m} P_j = \frac{\omega(1 - e^{-u})^s}{(s-1)!} \sum_{j \geq m} \frac{(j+s-1)! e^{-ju}}{j!} \tag{5.28}$$

(quantum model)

---

*There is something of a circular argument involved here, for if the activation energies were low, the reactants would be unstable at room temperature, and would be relatively unobtainable.

†This terminology is widely used, but it is something of a misnomer, for the quantity is an enthalpy, and not an energy, of reaction. In spectroscopic parlance, a dissociation energy D_0 is the energy difference between the products of dissociation, $AX + BY$, and the $A\!-\!X\!-\!Y\!-\!B$ molecule, with all species in their lowest vibrational states. For a discussion of the differences between these quantities, see S. W. Benson, *Journal of Chemical Education*, **42**, 502 (1965).

or

$$k_0 = \omega \int_{E_c}^{\infty} P(E) \, dE = \frac{\omega}{(s-1)!(RT)^s} \int_{E_c}^{\infty} e^{-E/RT} E^{s-1} \, dE \qquad (5.29)$$

(classical model)

Because of the difference between P_j and $P(E)$ at high energies (Fig. 5.4), the classical form overestimates k_0 by a large factor. Note that the only *rate constant* in these expressions is the collision frequency ω. The integral in Eq. (5.29) can be transformed to

$$\int y^{s-1} e^{-y} \, dy = y^{s-1} e^{-y} - (s-1) \int y^{s-2} e^{-y} \, dy$$

so that

$$k_0 = \frac{\omega}{(s-1)!} \left(\frac{E_c}{RT}\right)^{s-1} e^{-E_c/RT} + \text{higher terms} \qquad (5.30)$$

Because E_c/RT is usually large (20 to 40), the higher terms can be neglected, to a good approximation.

The corresponding activation energy E_0 is defined in the usual way:

$$E_0 = RT^2 \frac{d \ln k_0}{dT} = E_c - (s-1)RT + RT^2 \frac{d \ln \omega}{dT} \qquad (5.31)$$

The last term will contribute another factor of $\frac{1}{2}RT$ if the hard-sphere collisional expression is used. Aside from this factor, we note that

$$E_\infty - E_0 = (s-1)RT \qquad (5.32)$$

so that the experimental activation energy is expected to decrease as the pressure decreases. The physical significance of this point may best be understood by means of the Tolman expression (5.27). Since the average energy of all the molecules in the reaction system is unchanged by the pressure decrease, the average energy $\bar{\bar{E}}$ of reacting molecules must be lower by $(s-1)RT$ than it is at high pressures; hence it is simply $E_c + RT$. This results from the fact that energized molecules have less opportunity to lose energy by collision. Consequently, even those with energy just above E_c will ultimately react before they are de-energized. Or, another way of expressing this is to state that because of the strong energy dependence of k_E, the molecules with greatest energy will be removed more rapidly by decomposition than they can be formed by collision, thus skewing the distribution function toward lower energies. For the isomerization of methyl isocyanide, the experimental difference between values of E_a at the high- and low-pressure limits is 2 kcal/mole.

5.7. *The Efficiency of Various Gases in Activating and Deactivating Collisions*

Measurements of unimolecular rate constants at pressures sufficiently low for Eq. (5.28) or (5.29) to be applicable provide a method for evaluating *relative* efficiencies of various molecules for collisional energy transfer. If no foreign gas is added to the reactant, a plot of k_0 against (A) in this pressure region will give a straight line of slope $Z_{AA} \sum P_j$. For other added gases, the same procedure is used to obtain $Z_{AM} \sum P_j$, if (A) $\ll$ (M). Since the properties of the reactant molecule are independent of M, P_j is the same in both cases, and the relative efficiency ϕ_M is given by

$$\phi_M = \frac{(\text{slope})_M}{(\text{slope})_A} = \frac{Z_{AM}}{Z_{AA}} \tag{5.33}$$

If the concentration of A is not negligible with respect to that of M, an additivity relationship

$$\omega = Z_{AA}(A) + Z_{AM}(M) \tag{5.34}$$

is used to evaluate the relative efficiencies. This factor may be put on a "per collision" basis by correcting for bimolecular collision rates, so that

$$\phi'_M = \phi_M \left(\frac{\mu_{AM}}{\mu_{AA}}\right)^{1/2} \left(\frac{\sigma_{AA}}{\sigma_{AM}}\right) \tag{5.35}$$

Typical results are given in Fig. 5.7 and Table 5.4, where data for the reaction $CH_3NC + M \longrightarrow CH_3CN + M$ have been used. It is seen that relative efficiency tends to increase with increasing molecular weight and with increasing molecular complexity. Indeed, a strong correlation between collision efficiency and boiling point, which is related to the strength of intermolecular forces, was found. This same type of behavior is found in other unimolecular reactions.

Unfortunately, this type of experiment does not give absolute rate constants or evidence to test our assumption of "strong collisions." The study of the exchange of vibrational and translational energy is a large subject, of considerable importance not only in chemical kinetics but also in determining transport properties of gases.* Unfortunately, most of the theoretical and experimental work deals with vibrational relaxation from the first excited state to the ground state ($v = 1 \rightarrow 0$), whereas it is apparent from typical values of E_c that we are concerned

*For a recent review, see Reference 9.

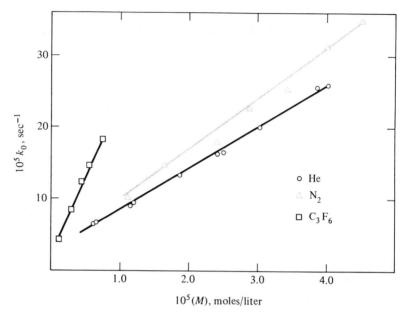

Figue 5.7. *Observed rate constant k_0 for the isomerization of methyl isocyanide in the low-pressure region. Added gases are: o, He; Δ, N_2; □, C_3F_8. The finite intercepts are an indication of a heterogeneous contribution to the observed rate.* [*Data from Rabinovitch, et. al.*, J. Phys. Chem. **74**, *3160 (1970).*]

Table 5.4. *Relative Efficiencies for De-energization of* CH_3NC[a]

Added gas	ϕ'_M	Added gas	ϕ'_M
He	0.243	NH_3	0.93
Ne	0.276	CD_3F	0.68
Ar	0.279	CH_3NC	(1.00)
Kr	0.245	CD_3CN	0.91
Xe	0.232	C_2H_5CN	0.88
H_2	0.24	C_3F_8	1.01
N_2	0.38	$n-C_6H_{14}$	0.99

[a]From Rabinovitch et al., *Journal of Physical Chemistry*, **74**, 3160 (1970). A total of 101 added gases were investigated in this work.

with transitions from a low-lying state to a highly excited state, or from a highly excited state to one below the critical energy.

Nonkinetic experiments give us the following information. For diatomic molecules, from 10^3 to 10^9 collisions are needed for a $v = 1 \rightarrow 0$ transition to take place, depending on the particular molecule and the temperature. For a given diatomic molecule, the transition probability is increased at higher values of v. Also, for polyatomic molecules, the $v = 1 \longrightarrow 0$ transition requires an average of only 10^2 to 10^4 collisions at room temperature. The combination of these facts makes it reasonable to hope that deactivation of polyatomic molecules in high vibrational states does not require many collisions.

Little direct experimental evidence bearing on this point is available. One source of information is the study of fluorescence enhancement by collisional deactivation of an excited electronic state. In the case of β–naphthylamine with an excess energy of 24 to 29 kcal/mole, at each collision He and H_2 remove a few tenths of a kilocalorie per mole, while polyatomic molecules de–energize the amine to the extent of a few kilocalories per mole at each collision (10). Schlag and co-workers (11) have revived this technique to measure absolute rates of de-energizing collisions and find that the rates are correlated with the extent of vibrational excitation but are roughly an order of magnitude below the rate for collisions.

Another approach is the method of "chemical activation" described in Section 5.11. Using this technique to produce sec-butyl radicals with 40 kcal/mole of excess vibrational energy, Kohlmaier and Rabinovitch (12) have determined the average amount of energy removed per collision. Depending on the inert gas added, this amount ranges from 1 to 3 kcal/mole (rare gases) to 9 kcal/mole (polyatomic molecules). Previously we stated that the average energy of molecules that react is $E_c + sRT$, so that an energy loss of sRT will, on the average, lead to deactivation. The results of Kohlmaier and Rabinovitch are in this range. Furthermore, the validity of the strong collision assumption is reinforced by the very sharp drop–off in k_E as a function of energy. Thus even if a molecule does not lose enough energy in a collision to drop below the critical energy, the loss of a few kilocalories per mole may lead to a tenfold lower decomposition rate (see Fig. 5.4).

5.8. Rate Constants at Intermediate Pressures

At pressures between the high- and low-pressure limiting regions, the expression for the rate constant becomes cumbersome. The quantum RRK model leads to the general expression

$$k_{uni} = D(1 - e^{-u})^s \sum_{j \geq m} \frac{C_j e^{-ju}}{1 + (D/\omega)(C_j/g_j)} \tag{5.36}$$

where

$$C_j = \frac{(j - m + s - 1)!}{(j - m)!(s - 1)!}$$

$$g_j = \frac{(j + s - 1)!}{j!(s - 1)!}$$

The analogous classical expression is

$$k_{uni} = \frac{De^{-y_c}}{(s - 1)!} \int_{E_c}^{\infty} \frac{(y - y_c)^{s-1} \exp(y_c - y)\, dy}{1 + (D/\omega)[1 - (y_c/y)]^{s-1}} \qquad (5.37)$$

which must be integrated numerically. This expression makes it clear that the pressure dependence is a function of D/ω, y_c, and s. The effect of s is shown in Fig. 5.8 in terms of a log-log plot of k_{uni}/k_∞ against the dimensionless parameter $\theta = \omega/D$. For calibration of this latter variable, note that if D is 10^{14} sec^{-1} and ω is calculated for hard-sphere collisions of molecules with atomic weight 50 and diameter 5 Å, then $\theta = 5 \times 10^{-8}$

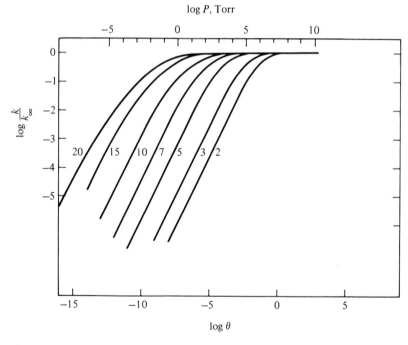

Figure 5.8. *Fall-off behavior of k_{uni} for various values of s (indicated on each curve.) The abscissa is $\theta = \omega/D$ at the bottom and at the top it is $P(Torr)$ for the case where $D = 10^{14}$ sec^{-1}, $M = 50$, $d = 5$ Å. Values of k_j were calculated from the Kassel quantum expression with $m = 35$, $u = 1$.*

at one Torr and 500 °K. The values of k_{uni}/k_∞ were obtained from Eq. (5.36) with $u = 1$ and $m = 35$. One notes that for small values of s, the first-order region is reached only at unattainably high pressures; conversely, the limiting second-order slope is found only in the high-vacuum region when s is large. Also, the drop-off of rate constant with decreasing pressure becomes more gradual as s increases.

The influence of the other variable, E_c/RT or mu, can be determined from Table 5.2, which indicates that an increase in y_c brings about a decrease in k_E/D, for a specific choice of s. The result will shift the curves of Fig. 5.8 toward lower values of θ.

The decrease in rate constant with decreasing pressure can be used to determine the rate constant k_j averaged over the steady-state distribution of energized molecules. Consider the quantity

$$\frac{k_\infty - k_{uni}}{k_{uni}} = \frac{\sum k_j P_j - \omega \sum k_j P_j/(\omega + k_j)}{\omega \sum k_j P_j/(\omega + k_j)} \tag{5.38}$$

obtained from Eqs. (5.7) and (5.10). Then

$$\frac{k_\infty}{k_{uni}} - 1 = \frac{\sum \left(\frac{k_j}{\omega}\right)\left(\frac{\omega k_j P_j}{\omega + k_j}\right)}{\sum \omega k_j P_j/(\omega + k_j)}$$

But since the second term in parentheses is the steady-state distribution function P'_j, the right-hand side of (5.38) is an average of k_j/ω over this particular energy distribution; that is,

$$\langle k_j \rangle P'_j = \omega \left(\frac{k_\infty}{k_{uni}} - 1\right) \tag{5.39}$$

Curves of the type shown in Fig. 5.8 can be used in conjunction with experimental data to determine empirically the appropriate value of s for a particular unimolecular reaction. (Usually this step is done with the analogous curves derived from the classical RRK formulation.) Values of E_c and k_∞ are determined from experiments in the limiting high-pressure region, and the results at lower pressures then allow one to search for the theoretical curve that is in best agreement. This procedure does not give an unambiguous choice of s, for it is evident that the shape of the curves is nearly the same for nearby values of s.

5.9. Other Forms of the Energy Distribution Function P(E)

The three important factors in the observable rate constant k_{uni} are k_E, the rate constant for decomposition at a specific energy, $P(E)$, the energy distribution function, and ω, the effective frequency of energy-removing collisions. We have already discussed this last factor in some detail.

The problem in the evaluation of $P(E)$ arises from the computation of the weighting factor $G(E)\,dE$ or the degeneracy g_j. It is useful to define the weighting factor in terms of the density of states at a given energy; this is

$$G(E) = \frac{dN(E)}{dE} \tag{5.40}$$

where $N(E)$ is the total number of states with energy between 0 and E. For a real molecule, it is a good approximation to express the total vibrational energy in terms of a combination of harmonic oscillator levels such that

$$E_v = h \sum_{i=1}^{3N-6} (v_i + \tfrac{1}{2}) v_i \tag{5.41}$$

where the v_i are quantum numbers and the v_i are vibrational frequencies. Then $N(E)$ is the number of different combinations of $v_1, v_2, \ldots, v_{3N-6}$ that, when substituted in (5.41), give a value of E_v no larger than the total energy E. The calculation of $N(E)$ according to this method is a task that only a large digital computer can tackle, except when all v_i are equal. In the latter case, the Kassel quantum form (Eq. 5.14) provides an exact count, and one can show that

$$N(E) = \sum_{j=0}^{j_{max}} g_j = \frac{(j_{max} + s)!}{j_{max}! s!} \tag{5.42}$$

where $E = j_{max} h v$. The classical formula (Eq. 5.19) can easily be extended to the general case when frequencies are not identical; it is

$$G(E) = \frac{E^{s-1}}{(s-1)! \Pi_i h v_i}$$

$$N(E) = \frac{E^s}{s! \Pi h v_i} \tag{5.43}$$

Some caution in the choice of the energy origin is required; since this formula is classical, the energy should be measured from the minimum of the vibrational potential. However, it is clear from Eq. (5.41) that $G(E)$ cannot be evaluated at energies below the zero-point energy

$$E_z = \left(\frac{h}{2}\right) \sum v_i$$

This point is explicitly shown by the modification of Eq. (5.42) suggested by Marcus and Rice (13), in which the variable E is measured from E_z:

$$G(E) = \frac{(E + E_z)^{s-1}}{(s-1)! \Pi h v_i}$$

$$N(E) = \frac{(E + E_z)^s}{s! \Pi h v_i} \tag{5.44}$$

Equation (5.44) does not completely straighten out $G(E)$, however, because E is still treated as a continuous and not a quantized variable.

Still another error is inherent in any method other than an exact count of the number of states. If E is less than $h\nu$ of the highest-frequency oscillator, this particular oscillator cannot serve as an energy reservoir and the total number of effective oscillators is less than s. As a result, s increases with increasing energy, contrary to the assumptions of the original model.

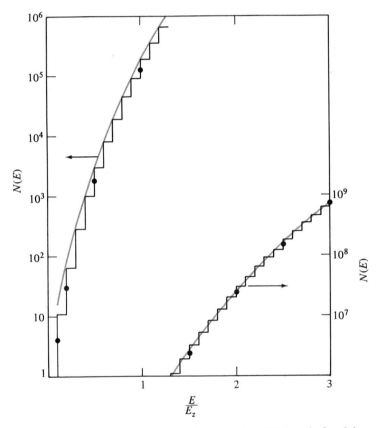

Figure 5.9. *Comparison of the number of states, $N(E)$, calculated from Eqs. (5.42), (5.44), and (5.45). A hypothetical molecule with 10 equal frequencies such that $h\nu = 0.1$ was used. Hence $E_z = 0.5$, and E is the energy in excess of this. The step function portrays $N(E)$ obtained by direct count, the solid smooth curve represents $N(E)$ from the Rice-Marcus expression, and the solid circles are values of $N(E)$ from the Whitten-Rabinovitch formula.*

A more sophisticated form of Eq. (5.44) has been recommended by Whitten and Rabinovitch (14), who use the expression

$$N(E) = \frac{(E + aE_z)^s}{s!\Pi h\nu_i} \tag{5.45}$$

where the parameter a is given by

$$a = 1 - \frac{w(s-1)}{s}\frac{\langle v^2 \rangle}{\langle v \rangle^2} \tag{5.46}$$

Equation (5.46) was applied to a number of actual molecules, and a was evaluated by a comparison of $N(E)$ calculated from the equation and from the exact count. The parameter w is strongly dependent on energy (approaching zero at large E) but does not change appreciably from molecule to molecule. Thus it is possible to calculate $N(E)$ for any actual molecule, with Eqs. (5.45) and (5.46) and the values of w as a function of E/E_z tabulated in Reference 14.

A comparison of $N(E)$ calculated according to the direct count method [Eq. (5.42)], the Rice-Marcus method [Eq. (5.44)], and the Whitten-Rabinovitch method [Eq. (5.45)] is given in Fig. 5.9. At energies slightly above the zero-point energy, the Rice-Marcus method overestimates $N(E)$ by a large factor, but this discrepancy disappears at higher energies.

We should point out that none of the foregoing methods of obtaining $N(E)$ has included the possible effect of anharmonic corrections to vibrational energy levels. One would expect these corrections to become significant at the relatively high energies of reacting molecules, and methods for taking this factor into account have been suggested. However, we shall not consider this refinement further.

5.10. Other Forms of the Rate Constant k_E

We would like to have a model for the rate constant k_E that is based on the properties of the reacting molecule and that does not require the introduction of ad hoc parameters such as D and s.

One such model was developed by Slater (15), who assumes that a reacting molecule is composed of harmonic oscillators incapable of exchanging energy with each other between collisions. Decomposition then occurs when some critical coordinate of the molecule (e.g., a bond length between two fissioning groups) reaches a limiting value. Since any molecular coordinate can be expressed in terms of the normal coordinates suitable for a description of molecular vibrations, the frequency with which the "critical distance" is reached can be expressed in terms of molecular vibration frequencies. The final expression for the unimolecular rate constant is very similar to the RRK expression, except that the factor D is now calculated a priori, from molecular frequencies. The number of effective molecular vibrations is generally found to be about twice the number of oscillators in the RRK model. The physical model used in this theory is an appealingly specific one, but the restriction on the intramolecular energy exchange (which distinguishes this theory from all others) appears unrealistic.

Another approach has been used by Rice and Marcus (13, 16), who have combined essential features of activated complex theory and RRK theory to produce what is commonly called the RRKM formulation of unimolecular rate constants. The important feature of this method is that it provides a means of evaluating the factor D, which appears somewhat arbitrarily in the original RRK theory.

The reaction scheme of Eq. (5.1) is expanded slightly to include an additional step:

$$A + M \underset{k_d}{\overset{k_a}{\rightleftharpoons}} A^* + M$$

$$A^* \xrightarrow{k_E} A^\ddagger \qquad\qquad\qquad\qquad (5.47)$$

$$A^\ddagger \xrightarrow{k_r} B$$

In these equations, A^* and $A^\ddagger$ are in a particular energy range. It is important to distinguish between A^*, the energized molecule with energy greater than the critical energy for reaction, and the activated complex $A^\ddagger$, which also has $E \geq E_c$ but, in addition, has some specific configuration. On the potential energy surface represented in Fig. 5.1, the energized molecule might, for example, have an energy of -3 units, well above the height of the pass separating A—B—C and A + BC. All configurations of A—B—C at this level are A^* molecules. At this same level within the pass is the $A^\ddagger$ species, with definite values of r_{AB} and r_{BC}. The RRKM procedure uses statistical mechanics to calculate the chance that an energized molecule will have the $A^\ddagger$ configuration, and it multiplies this probability by a frequency factor that represents the rate of crossing the boundary in phase space between $A^\ddagger$ and B.

If the steady-state conditions for A^* and $A^\ddagger$ are used in conjunction with Eq. (5.47), the expression for k_{uni} is found to be identical with that of Eq. (5.2) and the rate constant k_r drops out. Thus we return to the usual problem of evaluating k_E, $P(E)$, and ω. From the steady-state condition for A, we have

$$k_E(A^*) = k_r(A^\ddagger) = \frac{(A^\ddagger)}{2t^\ddagger} \qquad\qquad\qquad (5.48)$$

where $t^\ddagger$ is the time required for crossing the boundary between $A^\ddagger$ and B, and the factor of one-half selects only systems moving in the right direction. The ratio $(A^\ddagger)/(A^*)$ can be expressed by statistical mechanics in terms of distribution functions or densities of states:

$$\frac{(A^\ddagger)}{(A^*)} = \frac{P(E)_{A\ddagger}}{P(E)_{A^*}} = \frac{G(E^\ddagger)}{G(E^*)} \tag{5.49}$$

Note that the usual Boltzmann factor has dropped out, for both species are in the same energy level. Then, combining (5.48) and (5.49), we have for the rate constant at energy E

$$k_E = \frac{1}{2t^\ddagger} \frac{G(E^\ddagger)}{G(E^*)} \tag{5.50}$$

Remember that A^* is just the normal reactant molecule, except for its energy content. Hence we already know how to evaluate $G(E^*)$, the density of states of the energized molecule, using one of the available procedures discussed in the last section.

The factor $t^\ddagger$ is evaluated by a procedure which parallels that used in activated complex theory to obtain the frequency of passage through the surface dividing reactants from products (Section 4.4).

The density of states for the activated complex, $G(E^\ddagger)$, is more difficult to evaluate, and the details will not be given here. It is first necessary to factor the total density $G(E^\ddagger)$ into a term $G(E_t^\ddagger)$, which takes care of the kinetic energy of motion along the reaction coordinate, and a term $G(E^\ddagger - E_t^\ddagger)$ for all the other degrees of freedom. The details of the latter depend on the internal degrees of freedom (vibrations and internal rotations) that are "active"—that is, capable of exchanging energy.

A particularly simple result arises if rotational levels do not change during the process $A^* \rightarrow A^\ddagger$. In that case, the only contribution from rotational degrees of freedom to the rate constant is the ratio of partition functions

$$\left(\frac{Q_r^\ddagger}{Q_r^*} \right) = \frac{(I_A^\ddagger I_B^\ddagger I_C^\ddagger)^{1/2}}{(I_A I_B I_C)^{1/2}}$$

where the I's are moments of inertia. Then both $G(E^*)$ and $G(E^\ddagger - E_t^\ddagger)$ are entirely vibrational, the former for s active vibrations and the latter for only $s - 1$, since one vibration has become the reaction coordinate. If both state densities are calculated by means of Eq. (5.43), the resulting rate constant is

$$k_E = \left[\frac{Q_r^\ddagger}{Q_r} \frac{\prod_{i}^{s} \nu_i}{\prod_{i}^{s-1} \nu_i^\ddagger} \right] \left(\frac{E^\ddagger + E_z^\ddagger}{E + E_z} \right)^{s-1} \tag{5.51}$$

In this expression, the critical energy (E_c) has been used as the energy origin of the transition state, so that

$$E^\ddagger = E^* - E_c \tag{5.52}$$

Then one can see that Eq. (5.51) reduces to the classical RRK form, if the zero-point energies are set equal to zero. The hitherto undefined factor D is then given by the bracketed term in Eq. (5.51). Thus it depends on moments of inertia and vibrational frequencies of the reactant molecule and the activated complex. One should not think that all empiricism has thus been removed, for it is still necessary to guess how the frequencies change when the energized molecule becomes the activated complex. In particular, it is necessary to decide which vibration becomes the reaction coordinate.

For this same special case, the rate constant at the high-pressure limit can also be obtained, with the result

$$k_\infty = \frac{\mathbf{k}T}{h} \frac{Q_r^\ddagger Q_v^\ddagger}{Q_r Q_v} e^{-E_c/\mathbf{k}T} \tag{5.53}$$

where $Q_v^\ddagger$ and Q_v are vibrational partition functions. Equation (5.53) is merely the form that ordinary activated complex theory would produce.

This rate expression now provides us with a basis for discussing the Arrhenius pre-exponential factors of Table 5.3. In many unimolecular reactions, this factor is within the range of 10^{13} to 10^{15} sec^{-1}. The lower of these values is approximately the value of $\mathbf{k}T/h$ at 500 °K, which is toward the lower end of the temperature region in which unimolecular reactions are studied.

A lowering of vibrational frequencies in the activated complex or a conversion of vibrational degrees of freedom in the molecule into rotational motion in the activated complex will multiply this value by a factor greater than unity. Extreme examples are provided by several cases of fission to polyatomic radicals, where A factors as large as 10^{18} have been observed. These values may be explained if some vibrations of the molecule have become essentially free rotations in a very loose activated complex, so that small Q_v terms have become large Q_r terms. Conversely, the observation in a few cases of A factors as low as $10^{11.5}$ can be linked to a cyclic activated complex in which vibrational degrees of freedom have replaced internal rotations of the molecule, so that $Q_v^\ddagger Q_r^\ddagger / Q_v Q_r$ is small. An example is provided by

$$C_2H_5-O-CH=CH_2 \longrightarrow \begin{array}{c} \text{[cyclic structure]} \end{array} \longrightarrow C_2H_4 + CH_3CHO$$

More specific details of the RRKM method, when other forms of the state density function are used and when rotational degrees of freedom are not adiabatic, may be found in References 13, 16, and 17. In addition, these references contain numerous comparisons between experiment and RRKM theory that demonstrate the utility of this method.

5.11. Unimolecular Reactions
in Systems Not in Thermal Equilibrium:
The Method of Chemical Activation

The most sensitive test of unimolecular rate theory would be the direct determination of k_E as a function of energy. From Fig. 5.6 it is apparent that this quantity is not easily obtained from the measurement of a unimolecular rate constant in a gas at thermal equilibrium. The very rapid drop-off of $P(E)$ at energies where k_E becomes appreciable leads to an averaging of k_E strongly weighted toward the critical energy region.

Experimentally, one would like to produce energized molecules with a specific energy and measure the rate of their decomposition or isomerization. One possible approach is through photoexcitation to higher electronic states that are also vibrationally excited. However, experiments of this type require a more detailed knowledge of molecular energy levels and spectra than is usually available.

A completely different method has been extensively applied by Rabinovitch and co-workers (16, 17); this technique has been labeled "chemical activation". It may best be described by considering a specific example: the decomposition of *sec*-butyl radicals produced by the reaction of hydrogen atoms with *cis*-butene-2. The important steps of the reaction are

$$H + cis\text{--}CH_3CH\text{=}CHCH_3 \rightleftharpoons CH_3CH_2\dot{C}HCH_3^* \qquad (5.54)$$

$$CH_3CH_2\dot{C}HCH_3^* \xrightarrow{k_E} \dot{C}H_3 + CH_3CH\text{=}CH_2 \qquad (5.55)$$

$$CH_3CH_2\dot{C}HCH_3^* + M \xrightarrow{\omega} CH_3CH_2\dot{C}HCH_3 + M \qquad (5.56)$$

Reaction (5.54) is exothermic by 40 kcal/mole, and this energy appears as the vibrational excitation of the *sec*-butyl radical. Because this heat of reaction is considerably larger than the activation energy for the decomposition reaction (5.55), the average energy of the reacting butyl radicals is much higher than it would be in the same decomposition under conditions of thermal equilibrium. The energy distribution can also be shown to be narrower. Moreover, the average energy can be shifted slightly by the use of other butene isomers and by isotopic substitution.

The fate of energized radicals with respect to decomposition or collisional deactivation is determined by the relative rates of reactions

(5.55) and (5.56), together with secondary reactions that follow these reactions. The yield of products arising from reaction (5.55) may be designated D (for decomposition) and that of products from (5.56), S (for stabilization). In terms of the three fundamental quantities that always appear in a unimolecular rate constant, the ratio D/S takes the form

$$\frac{D}{S} = \frac{\int_{E_c}^{\infty} \left(\frac{k_E}{\omega + k_E}\right) f(E)\, dE}{\int_{E_c}^{\infty} \left(\frac{\omega}{\omega + k_E}\right) f(E)\, dE} \tag{5.57}$$

The term in parentheses represents the fraction of radicals with energy in a specified interval dE that undergo either decomposition or collisional stabilization, and the $f(E)\, dE$ term gives the fraction of all radicals within the required energy range.

Thus it becomes possible to relate the experimental quantity D/S to k_E, which can be calculated for various models of the activated complex, according to the RRKM method. This approach has been extended to other alkyl radicals and has provided numerous studies of

1. The effect of pressure on the ratio D/S.
2. The variation of k_E with energy.
3. Effects of isotopic substitution (and hence of changes in vibrational frequencies) on $G(E^*)$ and $G(E^\ddagger)$.
4. The process of energy transfer by collision.

References

1. F. A. Lindemann, *Transactions of the Faraday Society* **17**, 598 (1922).
2. F. W. Schneider and B. S. Rabinovitch, *J. Am. Chem. Soc.* **84**, 4215 (1962).
3. C. N. Hinshelwood, *Proceedings of the Royal Society (London)* **A113**, 230 (1926).
4. O. K. Rice and H. C. Ramsperger, *Journal of the American Chemical Society* **49**, 1617 (1927); **50**, 617 (1928). O. K. Rice, *Proceedings of the National Academy of Sciences of the U.S.* **14**, 114, 118 (1928).
5. L. S. Kassel, *Kinetics of Homogeneous Gas Reactions*, Van Nostrand Rheinhold, New York, 1932; gives a summary with earlier references.
6. R. C. Tolman, *J. Am. Chem. Soc.* **42**, 2506 (1920); *ibid.* **47**, 2652 (1925).
7. I. Oref and B. S. Rabinovitch, *Journal of Physical Chemistry* **72**, 4488 (1968).
8. J. A. Kerr, *Chemical Reviews* **66**, 465 (1966).
9. R. G. Gordon, W. Klemperer, and J. I. Steinfeld, *Annual Reviews of Physical Chemistry* **19**, 215 (1968).
10. M. Boudart and J. T. Dubois, *Journal of Chemical Physics* **23**, 223 (1955).

11. E. W. Schlag and H. von Weyssenhoff, *J. Chem. Phys.* **51**, 2508 (1969).
12. G. H. Kohlmaier and B. S. Rabinovitch, *J. Chem. Phys.* **38**, 1692, 1709 (1963).
13. R. A. Marcus and O. K. Rice, *Journal of Physical and Colloid Chemistry* **55**, 894 (1951); R. A. Marcus, *J. Chem. Phys.* **21**, 359 (1952).
14. G. Z. Whitten and B. S. Rabinovitch, *J. Chem. Phys.* **38**, 2466 (1963).
15. N. B. Slater, *The Theory of Unimolecular Reactions*, Cornell University Press, Ithaca, New York, 1959. This book not only explains Slater's approach in detail but is also a useful source for the RRK theory.
16. G. M. Wieder and R. A. Marcus, *J. Chem. Phys.* **37**, 1835 (1962); R. A. Marcus, *J. Chem. Phys.* **43**, 2658 (1965).
17. B. S. Rabinovitch and M. C. Flowers, *Quarterly Reviews of the Chemical Society* (*London*) **18**, 122 (1964); B. S. Rabinovitch and D. W. Setser, *Advances in Photochemistry* **3**, 1(1964); contain references to chemical activation work done prior to 1964.

Problems

5.1. The first-order rate constant for the thermal decomposition of N_2O_5 in the presence of NO, defined by

$$-\frac{d(N_2O_5)}{dt} = k(N_2O_5)$$

varies with total pressure. In the presence of large amounts of nitrogen at 27 °C, the data are [R. L. Mills and H. S. Johnston, *J. Am. Chem. Soc.* **73**, 938 (1951)]

P, Torr	k, sec^{-1}	P, Torr	k, sec^{-1}
7000	0.265	475	0.116
3900	0.247	440	0.108
2400	0.248	400	0.104
2300	0.223	334	0.092
		300	0.086

(*a*) If the reaction is interpreted in terms of the mechanism

$N_2O_5 \longrightarrow NO_2 + NO_3$ slow

$NO + NO_3 \longrightarrow 2NO_2$ fast

which step does k refer to?

(*b*) Estimate k_∞ by using the Lindemann mechanism.

(*c*) In the absence of NO, there is no pressure dependence of the thermal decomposition of N_2O_5 in this pressure region. Ogg suggested the mechanism

$$M + N_2O_5 \rightleftharpoons NO_2 + NO_3 + M$$

$$NO_2 + NO_3 \longrightarrow NO_2 + NO + O_2$$

$$NO + NO_3 \longrightarrow 2NO_2$$

How does this mechanism explain the lack of pressure dependence?

5.2. The photolysis of ozone may be interpreted in terms of the mechanism

$$O_3 + h\nu \longrightarrow O_2 + O$$

$$O + O_2 + M \xrightarrow{k_1} O_3 + M$$

$$O + O_3 \xrightarrow{k_2} 2O_2$$

E. Castellano and H. J. Schumacher [*Zeitschrift fur Physikalische Chemie* **34,** 198 (1962)] find $10^3 k_1/k_2$ (in units of M) to be 7.0, 2.4, 1.8, 2.7, 3.1, and 6.7 when M is O_3, He, Ar, N_2, O_2, and CO_2, respectively. What are the relative efficiencies of these gases for activating ozone in the unimolecular reaction

$$O_3 \longrightarrow O_2 + O?$$

5.3. Use Tolman's theorem, Eq. (5.27), to show that the activation energy of a unimolecular reaction at the high-pressure limit is E_c. Use the classical form of k_E and $P(E)$.

5.4. (*a*) What value of A_∞ corresponds to an entropy of activation of 0 cal deg^{-1} mole^{-1}?

(*b*) For comparison, calculate $\Delta S^{o\ddagger}$ for the decomposition of C_2H_6 (Table 5.3), a reaction involving a "loose" activated complex.

(*c*) What is $\Delta S^{o\ddagger}$ for the decomposition of C_2H_5—O—CH=CH$_2$, which involves a cyclic activated complex?

5.5. Consider the following reaction system described by Setser and Rabino-vitch [*Can. J. Chem.* **40,** 1425 (1962)]. Ketene (CH_2CO) is photolyzed at 3200 Å in a large excess of ethylene at 300 °K. The methylene radicals thus produced have 21 kcal/mole total energy. The reaction sequence that follows is

$$CH_2 + CH_2=CH_2 \longrightarrow \nabla^* \qquad \Delta H = -79 \text{ kcal/mole}$$

$$\nabla^* + C_2H_4 \longrightarrow \nabla + C_2H_4$$

$$\nabla^* \xrightarrow{k_E} CH_3—CH=CH_2$$

(*a*) Use the data below to calculate k_E at the appropriate energy on the basis of Eq. (5.51).

(*b*) What would be the value of A in the RRK formulation with $s = 10$ that would give this same k_E?

(*c*) Using the hard-sphere model for deactivating collisions, calculate the ratio of cyclopropane to propene formed at a pressure of 400 Torr. (Assume that the methylene radicals are monoenergetic.)

Data: $E_a = 62.7$ kcal/mole for cyclopropane isomerization to propene (at infinite pressure).

	Active molecule	Activated complex
$\lvert I \rvert^{1/2}$	1.00	1.21
ν's	3050 (6) (6 identical frequencies)	3020 (5)
	1460 (3)	1430 (3)
	1080 (7)	1100 (4)
	870 (3)	900 (5)
	740 (2)	550 (3)

Collision diameters: Ethylene, 5 Å, Cyclopropane, 5 Å.

6

Reactions

in

Solution

THE NATURE OF THE SOLVENT has no effect on some reaction rates and a large effect on others. The rates of unimolecular reactions and reactions between nonpolar species are often independent of the composition of the solvent, whereas the rates of reaction of ions are strongly dependent on the solvent. In fact, gas-phase reactions at normal temperatures seldom involve ionic species unless the ions are produced by high-energy processes (such as electron bombardment in the source of a mass spectrometer). The reason is evident from an estimate of the ionization energies of neutral species, such as HCl, in the gas phase and in water:

$$HCl(g) \longrightarrow H^+(g) + Cl^-(g) \qquad \Delta H^\circ = 327 \text{ kcal/mole}$$

$$HCl(aq) \longrightarrow H^+(aq) + Cl^-(aq) \qquad \Delta H^\circ = -25 \text{ kcal/mole}$$

The difference between these two values results from very large negative enthalpies of solution for the two ions, compared with a small negative enthalpy of solution for HCl.

Some solvent effects are due to long-range forces acting over distances of several angstroms or more. In bimolecular reactions, they are largely manifested by concentration gradients of one reactant around the other. Such effects are the simplest to understand because the detailed structure of the solvent need not be considered.

6.1. Diffusion and Conduction

In Chapter 3 the rate constant for a gas-phase reaction proceeding on every collision was shown to be the product of the relative velocity of the reactants and a collision cross section. This result is equally applicable to reactants in well-ordered beams or to reactants undergoing random motion in a vessel at atmospheric pressure. Although motion of gas molecules at the higher pressure is impeded by collisions between molecules, the resulting random motion (or diffusion) does not change the collision rate because the average distance between collisions is much larger than the radii of the reactants.

The opposite is true in liquids. It is hard to define precisely a mean free path in liquids because each molecule always interacts strongly with at least one neighboring molecule. However, in view of the small amount of free space in a liquid, it is clear that mean free paths are small compared to molecular dimensions. Consequently, diffusion effects are important in liquid-phase reactions.

If a solute has a higher concentration in one region of a solution than in another, it will tend to diffuse toward the region of lower concentration. This fact is expressed quantitatively by Fick's first law of diffusion:

$$\boldsymbol{\phi}_i = -D_i \nabla C_i \qquad (6.1)$$

where $\boldsymbol{\phi}_i$ is the flux of the solute i at some point, expressed in units of (quantity) $\text{cm}^{-2}\text{sec}^{-1}$, ∇C_i is the concentration gradient in units of (quantity) $\text{cm}^{-3}\text{cm}^{-1}$, and D_i is the *diffusion coefficient* of the solute in the particular solvent. The units of D_i are $\text{cm}^2 \text{ sec}^{-1}$, and one notes that the units of quantity in which the flux and concentration gradient are expressed cancel. The magnitude of D_i is typically between 10^{-6} and $10^{-4} \text{ cm}^2 \text{ sec}^{-1}$, for small molecules in nonviscous solvents.

If the concentration gradient is spherically symmetrical—that is, if the concentration at a distance r from an origin depends only on r—then Eq. (6.1) becomes

$$\phi_i = -D_i \left(\frac{dC_i}{dr} \right) \tag{6.2}$$

The sign is chosen so that D_i is positive.

Diffusion is the process of completely random motion of the solute. Other forces, such as a potential gradient, can cause a net motion of the solute in a preferred direction. The flux of a solute at some point can be expressed as the product of a velocity and the concentration of the solute at that point (which is essentially another definition of flux):

$$\phi_i = \mathbf{v}_i C_i \tag{6.3}$$

We can define a mobility, u, for an ion moving in a potential gradient, as the ratio of the speed with which the ion moves to the potential gradient, ∇E.

$$u_i = \frac{\mathbf{v}_i}{\nabla E}$$

and from (6.3)

$$\phi_i = \frac{-z_i}{|z_i|} u_i C_i \nabla E \tag{6.4}$$

where the factor $-z_i/|z_i|$ accounts for the direction of motion of the ion—that is, a negative ion will have a positive flux in the direction of increasing E, whereas a positive ion will move in opposition to the field. The units of ∇E and u_i are usually volts cm^{-1} and cm^2 volt^{-1} sec^{-1}. If the concentration and potential gradient are spherically symetrical about an origin, Eq. (6.4) may be written

$$\phi_i = \frac{-z_i}{|z_i|} u_i C_i \frac{dE}{dr} \tag{6.5}$$

The utility of Eqs. (6.4) and (6.5) is due to the fact that the mobilities of ions in liquids are independent of the voltage gradient, since the energy gained by the ions between collisions is very small compared to normal kinetic energies and is rapidly dissipated by subsequent collisions.

Consider a conductance cell with two electrodes, each one cm^2 in area. The current carried by ion i is proportional to the flux of ion i; and if the concentration of i is expressed in moles per cm^2, then the proportionality constant will be the charge carried per mole of ions, or $z_i F$, where F is the Faraday constant, 96,487 coulombs per equivalent. From Eq. (6.4),

$$\mathbf{i} = |z_i| F u_i C_i \nabla E \tag{6.6}$$

It is more usual to express the current carried by an ion in a conductance cell in terms of the equivalent conductance of the ion, λ_i, which is equal to the current carried by one gram-equivalent of i in a potential gradient of one volt per centimeter.

$$\mathbf{i} = |z_i| \lambda_i C_i \nabla E \tag{6.7}$$

Comparing Eqs. (6.6) and (6.7),

$$u_i = \frac{\lambda_i}{F} \tag{6.8}$$

The total current in the conductance cell is equal to the sum of all the individual ion currents. The separate equivalent conductances are determined by their additive nature combined with a transference number experiment for one electrolyte, in which the concentration changes produced by the ion fluxes are measured.

The actual motion of an ion in a liquid is due to both diffusion and conduction. The complete equation governing the movement is

$$\phi_i = -\left(D_i \frac{dC_i}{dr} + \frac{z_i}{|z_i|} u_i C_i \frac{dE}{dr} \right) \tag{6.9}$$

Diffusion is completely negligible in a real conductivity experiment, but this fact is not true for a chemical reaction. In a reaction between two negative ions, the ions will tend to move away from each other because of their mobility in the mutual field. Only diffusion brings the reactants together; thus for reaction to occur, the diffusion term in Eq. (6.9) must exceed the ionic mobility term.

Equation (6.9) may be simplified because of a relation between D and u. Consider a system at equilibrium in the presence of an electrostatic field. The establishment of equilibrium requires that the net flux of ions must be zero at all points in the solution:

$$\phi_i = -D_i \frac{dC_i}{dr} - \frac{z_i}{|z_i|} u_i C_i \frac{dE}{dr} = 0$$

$$\frac{dC_i}{C_i} = -\frac{z_i}{|z_i|} \frac{u_i}{D_i} dE$$

and

$$\ln \frac{C_i}{C_i^\circ} = -\frac{z_i}{|z_i|} \frac{u_i}{D_i} E \tag{6.10}$$

where C_i° is the concentration of ions at some arbitrary point at which $E = 0$. We also know that at equilibrium the ions in the potential field will have a distribution given by Boltzmann's law (Eq. 4.6):

$$C_i = C_i^\circ \exp \left[-\frac{V(r)}{kT} \right] \tag{6.11}$$

The potential energy of the ions, $V(r)$, is related to the potential by $z_i eE$, so Eq. (6.11) now becomes

$$C_i = C_i^0 \exp\left(-\frac{z_i e E}{kT}\right)$$

or

$$\ln \frac{C_i}{C_i^0} = \frac{-z_i e E}{kT} \tag{6.12}$$

Comparison of Eqs. (6.10) and (6.12) shows that

$$\frac{u_i}{D_i} = \frac{|z_i|e}{kT} \tag{6.13}$$

or*

$$\frac{u_i}{D_i} = 38.9\,|z_i| \qquad \text{at } 25°C$$

Combining (6.13) with (6.9), we have

$$\phi_i = -D_i\left(\frac{dC_i}{dr} + \frac{z_i e C_i}{kT}\frac{dE}{dr}\right)$$

or

$$\phi_i = -D_i\left[\frac{dC_i}{dr} + \frac{1}{kT}C_i\frac{dV(r)}{dr}\right] \tag{6.14}$$

6.2. Rapid Reactions

A large fraction of the reactions used by a chemist are very rapid—so rapid that they have been referred to as "instantaneous" in some textbooks. A large variety of techniques are available for studying these reactions; some of the more generally useful ones, such as the stop-flow method, the temperature jump method, flash photolysis, pulse radiolysis, NMR line broadening, sound absorption, and the rotating sector method are described in Appendix A. Several of these techniques allow reaction times as short as 10^{-10} sec and second-order rate constants as large as 10^{12} M^{-1} sec^{-1} to be measured. Two species at concentrations of 10^{-2} M reacting with a rate constant of only 10^3 M^{-1} sec^{-1} would have a first half-life of only one-tenth of a second, which may well be described as instantaneous in terms of normal laboratory practice.

*The evaluation of the constant is

$$\frac{e}{k} = \frac{4.80 \times 10^{-10}\text{esu}}{300 \times 1.38 \times 10^{-16}} \text{ erg/deg}$$

where the factor 300 is introduced to convert from cgs units to practical units of voltage. The units of u_i are cm^2 volt^{-1} sec^{-1} and of D_i are cm^2 sec^{-1}.

As implied earlier, the maximum rate of a reaction in solution is limited by diffusion. Consider a bimolecular reaction

$$A + B \longrightarrow products \tag{6.15}$$

where A and B are either neutral species or ions. We will assume that the reaction is so fast that whenever A and B are in contact they react. The experiment is started with only A in solution. A small concentration of B is suddenly produced by some means, such as photolysis or thermal dissociation due to a temperature jump. Initially these reactants are distributed uniformly throughout the solution. Let us consider the distribution of B around an individual A. If B is formed in actual contact with A, the two will react immediately. If they are initially separated by a few solvent molecules, they will probably encounter each other after a very short time and will react. After some time has elapsed. A will have collided with most of the molecules within the distance of a few molecular diamters. Evidently if A still remains, none of these surrounding molecules could have been B. Therefore the concentration of B in the vicinity of A must be considerably smaller than the average concentration throughout the solution, (B). Of course, a single A is too small a unit to treat statistically, but one can speak of the average concentration gradient at some distance from each A. We can apply Eq. (6.14), using r as the distance from A; the motion of this coordinate system can be accounted for by using $D_A + D_B$ as the diffusion coefficient. Thus

$$\phi_B = -(D_A + D_B)\left[\frac{d(B)_r}{dr} + \frac{(B)_r}{kT}\frac{dV(r)}{dr}\right] \tag{6.16}$$

where $(B)_r$ is the local concentration of B at a distance r from A.

The concentration gradient will approach a steady state in which the rate of diffusion toward A will equal the rate of reaction. The rate of diffusion toward a single A is the product of the surface area of a sphere around A and the negative of the flux, $-4\pi r^2\phi_B$, so that the total rate of reaction is

$$k(A)(B) = -(A)4\pi r^2\phi_B$$

and from Eq. (6.16)

$$k(B) = 4\pi r^2(D_A + D_B)\left[\frac{d(B)_r}{dr} + \frac{(B)_r}{kT}\frac{dV(r)}{dr}\right] \tag{6.17}$$

The right-hand side of Eq. (6.17) is a constant at all radii, for we have assumed that a steady-state concentration gradient has been obtained. Now we can find $(B)_r$ as a function of radius and reaction rate by integration of Eq. (6.17). The integration is simplified by the substitution

$$(B)_r^* = (B)_r \exp\left[\frac{V(r)}{kT}\right] \tag{6.18}$$

from which

$$\frac{d(B)_r^*}{dr} = \left[\frac{d(B)_r}{dr} + \frac{(B)_r}{kT}\frac{dV(r)}{dr}\right] \exp \frac{V(r)}{kT}$$

and Eq. (6.17) becomes

$$k(B) = 4\pi r^2(D_A + D_B)e^{-V(r)/kT}\frac{d(B)_r^*}{dr}$$

$$\frac{d(B)_r^*}{dr} = \frac{k(B)}{4\pi(D_A + D_B)}r^{-2}e^{V(r)/kT} \tag{6.19}$$

We will call the radius at which A and B are in contact R. Integrating between the limits of R and infinity, we have

$$(B)_\infty^* - (B)_R^* = \frac{k(B)\beta^{-1}}{4\pi(D_A + D_B)} \tag{6.20}$$

where

$$\beta^{-1} = \int_R^\infty e^{V(r)/kT}r^{-2}\,dr \tag{6.21}$$

The constant β has the dimensions of a length and the properties of a reaction radius. Removing the substitution (6.18) from Eq. (6.20), we have

$$(B)_\infty e^{V(\infty)/kT} - (B)_R e^{V(R)/kT} = \frac{k(B)\beta^{-1}}{4\pi(D_A + D_B)}$$

At large values of r, $V(r)$ approaches zero and $(B)_\infty$ is (B). (Although not rigorously true, it is an excellent approximation in all cases.) With this assumption,

$$(B)_R = (B)e^{-V(R)/kT}\left[1 - \frac{k}{4\pi\beta(D_A + D_B)}\right] \tag{6.22}$$

In the original discussion of reaction (6.15), we postulated that the reaction was so fast that A and B reacted immediately upon encountering each other. In this case, the concentration of B at the surface of A would be zero and the preceding equation could be solved for the rate of the reaction. This assumption is unnecessarily restrictive, however, and we shall consider instead the case in which A and B in contact react with a finite rate constant k_R. The rate of reaction, expressed in terms of $(B)_R$ is

$$-\frac{d(B)}{dt} = k(A)(B) = k_R(A)(B)_R \tag{6.23}$$

and by (6.22)

$$k(A)(B) = k_R(A)(B)e^{-V(R)/kT}\left(1 - \frac{k}{4\pi\beta(D_A + D_B)}\right)$$

$$k = \frac{4\pi(D_A + D_B)\beta}{1 + [4\pi(D_A + D_B)\beta/k_R e^{-V(R)/kT}]}$$

The units of the rate constant are cm^3 $molecule^{-1}$ sec^{-1}, and if k and k_R are expressed in units of M^{-1} sec^{-1},

$$k = \frac{k_D}{1 + [k_D/k_R e^{-V(R)/kT}]}$$ (6.24)

where (N_0 is Avogadro's number)

$$k_D = 4\pi(D_A + D_B)\beta\frac{N_0}{1000}$$ (6.25)

If the reaction is between two like species,

$$2A \longrightarrow products$$

the rate constant would be defined as (Section 1.2)

$$-\frac{d(A)}{dt} = 2k(A)^2$$

and

$$k_D = 4\pi D_A \beta\frac{N_0}{1000}$$ (6.26)

From Eq. (6.24) we see that k cannot exceed k_D no matter how large k_R is. For this reason, k_D is called the diffusion-limited rate constant.

6.3. Diffusion-Limited Reactions

The rate constants for diffusion-limited reactions can be predicted from Eq. (6.25) with the additional knowledge of R and the form of the potential function. We will first consider reactions in which A and B are uncharged. In this case, $V(r)$ is zero (dipole-dipole interactions are negligible). From Eq. (6.21)

$$\beta^{-1} = \int_R^\infty r^{-2}\, dr = R^{-1}$$

and from (6.25)

$$k_D = 4\pi(D_A + D_B)R\frac{N_0}{1000}$$ (6.27)

or, if the reactants are the same species,

$$k_D = 4\pi D_A R\frac{N_0}{1000}$$ (6.28)

Note that k_D is proportional to R, whereas collision-controlled reactions in the gas phase were found to depend on the cross section, πR^2 (Eq. 4.28)

A comparison of some rate constants for rapid reactions with those computed from Eqs. (6.27) and (6.28) is given in Table 6.1. Some of the diffusion constants are estimated by comparison with similar stable species. Note that only a few of the reactants are spherically symmetrical, although this was one of the assumptions of the derivation. The worst case from this point of view is the reaction between two ethyl radicals. The radius chosen for the reaction is related to the radius of a CH_2 group, and not the whole radical. For the other reactions, the reduced symmetry is not serious and an average van der Waals radius is used. None of the observed rate constants exceeds the diffusion limit by more than experimental error (which is 10 percent or more).

The assumption that no long–range interaction is operating on the reactants would seem to be violated by the last reaction in the table, $H^+ + NH_3 \longrightarrow NH_4^+$, in which an ion-dipole interaction is present. However, the high dielectric constant of water (78.5 at 25 °C) reduces the contribution to the rate constant of the electrostatic potential to about 10 percent.

Diffusion-limited reactions between ions

$$A^{z_A} + B^{z_B} \longrightarrow \text{products} \tag{6.29}$$

can also be discussed in terms of Eq. (6.25). The potential energy is given by

$$V(r) = \frac{z_A z_B e^2}{\mathscr{D} r} \tag{6.30}$$

where e is the charge on the electron, 4.80×10^{-10} esu, and $\mathscr{D}$ is the dielectric constant. From Eq. (6.21)

$$\beta^{-1} = \int_R^\infty \exp\left(\frac{z_A z_B e^2}{\mathscr{D} r k T}\right) r^{-2} \, dr$$

$$\beta = \frac{-z_A z_B r_0}{1 - \exp\left(z_A z_B r_0 / R\right)} \tag{6.31}$$

where

$$r_0 = \frac{e_2}{\mathscr{D} k T} \tag{6.32}$$

In water at 25 °C, r_0 is 7.1×10^{-8} cm. Note that the reaction radius β is positive for any combination of signs for the charges. If oppositely charged ions are reacting, the exponential in the denominator is small because R is generally less than r_0, and β is approximately $-z_A z_B r_0$. If the ions have the same charge, the exponential is considerably greater than unity and β becomes quite small for small R.

The rate constant k_D may be found by inserting Eq. (6.31) in Eq. (6.25):

$$k_D = -\frac{4\pi(D_A + D_B)z_A z_B r_0 N_0}{1000[1 - \exp\left(z_A z_B r_0 / R\right)]} \tag{6.33}$$

Table 6.1. *Diffusion-Limited Reactions*

Reaction	Solvent	T, °C	$D_A \times 10^5$	$D_B \times 10^5$	R	$k_{calc} \times 10^{-10}$	$k_{obs} \times 10^{-10}$
$I + I \rightarrow I_2$[a]	CCl_4	25	4.2	–	4	1.3	0.82
$OH + OH \rightarrow H_2O_2$[b]	H_2O	24	2.6	–	3	0.6	0.5
$OH + C_6H_6 \rightarrow C_6H_6OH$[b]	H_2O	24	2.6	1.1	3	0.8	0.33
$C_2H_5 + C_2H_5 \rightarrow C_4H_{10}$[b]	C_2H_6	−177	0.7	–	2	0.1	0.02
$H^+ + NH_3 \rightarrow NH_4^+$[c]	H_2O	25	10.0	3.	4	4.0	4.3

Units: D_A and D_B in $cm^2 \ sec^{-1}$, R in angstroms, k in $M^{-1} \ sec^{-1}$.

[a] Reference 6.

[b] L. M. Dorfman and M. S. Matheson, in *Progress in Reaction Kinetics*, G. Porter (Ed.) Pergamon Press, Oxford 1965, vol. 3, p. 237.

[c] M. Eigen et al., *ibid.*, vol. 2, p. 285.

Many diffusion-limited reactions between ions have been studied in aqueous solution, and a representative selection of rate constants is given in Table 6.2, together with values computed from (6.32) using a uniform value of 5×10^{-8} cm for R. Note that the excellent agreement between theory and experiment for reactions between oppositely charged ions fades seriously for ions of the same charge. In the latter case, the computed constant is very sensitive to the choice of R.

Table 6.2. *Diffusion-Limited Reactions Between Ions in Water (about 25 °C)*[a]

Reaction	$k_{obs} \times 10^{-10}$	$k_{calc} \times 10^{-10}$
$e_{aq} + Co(NH_3)_6^{3+}$	8.2	9.3
$e_{aq} + Cr(en)_3^{3+}$	7.8	9.3
$e_{aq} + Cr(H_2O)_6^{2+}$	6.2	6.5
$H^+ + SO_4^{2-}$	10.	6.5
$e_{aq} + Ag^+$	3.2	4.1
$OH^- + H^+$	14.	11.
$HS^- + H^+$	7.5	9.
$e_{aq} + NO_2^-$	0.46	0.97
$e_{aq} + NO_3^-$	0.85	0.97
$e_{aq} + SeO_4^{2-}$	0.10	0.38
$e_{aq} + Ni(CN)_4^{2-}$	0.41	0.38
$e_{aq} + Fe(CN)_6^{3-}$	0.30	0.13
$H^+ + (H_2NC_2H_4)_3NH^{3-}$	0.56	0.26

[a]From M. Anbar and E. J. Hart, in *Radiation Chemistry I*, E. J. Hart (Ed.), Advances in Chemistry Series, American Chemical Society, Washington, D.C., 1968, p. 79, and from Reference c, Table 6.1.

Equations (6.27) and (6.33) do not predict that the reactions in Tables 6.1 and 6.2 will be diffusion limited. They only give the upper limit of the rate constant. The reactions were found to agree reasonably well with the upper limit; hence they are said to be diffusion limited. All reactions of hydrogen ions with strong bases that have been studied appear to be diffusion limited. These rates are generally measured by relaxation methods (see Appendix A). The hydrated electron, which is essentially the conjugate base of the hydrogen atom, is produced in the flash photolysis of many inorganic aqueous solutions and in the pulse radiolysis of water. Both techniques are described in Appendix A. The majority of rate constants measured for hydrated electron reactions are diffusion limited.

If a recombination reaction

$$A + B \longrightarrow C \qquad (6.34)$$

is diffusion limited in the forward direction, the reverse dissociation reaction will also exhibit the effects of diffusion. The products A and B are formed close to each other and have a large probability of encountering each other and reforming C instead of escaping. Such effects are manifested as low-quantum yields in such reactions as the photolysis of azomethane in solution (1).

6.4. Time Dependence of Diffusion

The diffusion of solutes was discussed in Section 6.2 for the situation in which the flux was independent of time. Equations (6.1) or (6.14) are inadequate if the solute concentration is changing rapidly with time. A time-dependent form of the diffusion equation is needed. We will consider only the case in which $V(r)$ is zero.

Consider the volume between two spherical shells with radii r and $r + \Delta r$, as in Fig. 6.1. The net rate at which the solute molecules enter the volume is measured at radius r, for the flux is measured toward increasing r. This net rate is given by

$$\text{Rate entering} = 4\pi r^2 \phi_i$$

where $4\pi r^2$ is the surface area of the spherical shell at r. The net rate leaving the volume element at $r + \Delta r$ is given by

$$\text{Rate leaving} = 4\pi r^2 \phi_i + \Delta(4\pi r^2 \phi_i)$$

and the rate of accumulation of solute in the volume, $\partial N_i / \partial t$, is the difference between the two:*

$$\frac{\partial N_i}{\partial t} = -4\pi \, \Delta(r^2 \phi_i) \tag{6.35}$$

The rate at which the concentration of the solute changes with time in this volume is found by dividing Eq. (6.35) by the volume, which for small values of Δr is $4\pi r^2 \, \Delta r$.

$$\frac{\partial C_i}{\partial t} = \frac{-\Delta(r^2 \phi_i)}{r^2 \, \Delta r}$$

In the limit, as $\Delta r \to 0$,

$$\frac{\partial C_i}{\partial t} = \frac{D_i}{r^2} \frac{\partial}{\partial r}\left(r^2 \frac{\partial C_i}{\partial r}\right) \tag{6.36}$$

The differential on the right-hand side of the equation is the Laplacian of C_i, and the general form of Eq. (6.36) is

---

*Note that the right-hand side of Eq. (6.35) is not the same as $-4\pi r^2 \, \Delta \phi_i$. The flux could be constant and N_i would still change with time because the surface area is larger at $r + \Delta r$.

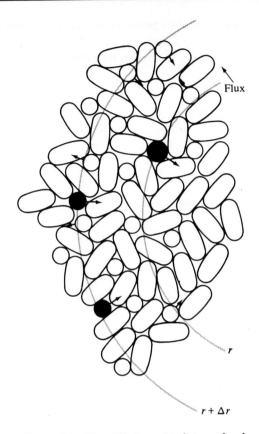

Figure 6.1. *The diffusion of solute molecules (circles) in a solvent. The concentration of solute increases with radius so that there is a net movement of solute towards the origin, indicated by the filled circles.*

$$\frac{\partial C_i}{\partial t} = D_i \nabla^2 C_i \qquad\qquad (6.37)$$

which is Fick's second law of diffusion.

Two conditions for a steady-state solution of Eq. (6.36) are apparent. The first is that $\partial C_i/\partial t = 0$ when $\partial C_i/\partial r = 0$—that is, when the concentration is uniform throughout the solution. The second is that $\partial C_i/\partial t = 0$ when

$$\frac{\partial C_i}{\partial r} = \frac{\text{constant}}{r^2}$$

which is the solution to be found from Eq. (6.19) if $V(r)$ is zero.

163

We shall determine the most probable position of a particle at time t after starting from $r = 0$ at $t = 0$. One cannot speak of the concentration of a single particle, but Eq. (6.36) can be written as

$$\frac{\partial w}{\partial t} = \frac{D}{r^2} \frac{\partial}{\partial r}\left(r^2 \frac{\partial w}{\partial r}\right) \tag{6.38}$$

where w is the probability per unit volume of finding the particle at radius r. Diffusion is a process of Brownian motion, which suggests that w will have the form of a gaussian function:

$$w = (4\pi b^2)^{-3/2} \exp\left(\frac{-r^2}{b^2}\right) \tag{6.39}$$

where b^2 is a time-dependent parameter. By differentiation,

$$\frac{\partial w}{\partial t} = \frac{w}{b^2}\left(\frac{r^2}{b^2} - \frac{3}{2}\right)\frac{\partial b^2}{\partial t}$$

$$r^{-2}\frac{\partial}{\partial r}\left(r^2\frac{\partial w}{\partial r}\right) = \frac{4w}{b^2}\left(\frac{r^2}{b^2} - \frac{3}{2}\right)$$

and introducing these derivatives into Eq. (6.38), we find

$$\frac{\partial b^2}{\partial t} = 4D \tag{6.40}$$

This result demonstrates that Eq. (6.39) is the required solution because Eq. (6.40) is independent of r. Integration of (6.40) gives

$$b^2 = 4Dt \tag{6.41}$$

From Eq. (6.39) we see that b is the radius at which w has decreased to a factor of $1/e$ of its maximum value, but it is also the most probable radius for finding the particle, which may be verified by searching for the maximum in the function $4\pi r^2 w$. Equation (6.41) is useful in estimating the time required for a particle to move a given distance. If $4D$ is 10^{-4} cm^2 sec^{-1}, a typical value, a particle will diffuse one centimeter in 3 hours.

We shall use Eq. (6.41) to estimate the time required to establish the steady-state concentration gradient in a diffusion-limited reaction. Integration of Eq. (6.19) between the limits r and ∞ $[V(r) = 0]$ leads to

$$(B)_r = (B)\left\{1 - \frac{k}{4\pi r(D_A + D_B)}\right\} \tag{6.42}$$

If the reaction is diffusion limited, the rate constant k is given by Eq. (6.27), or, in units of molecules cm^{-3} sec^{-1},

$$k = 4\pi(D_A + D_B)R$$

and with (6.42)

$$(B)_r = (B)\left(1 - \frac{R}{r}\right)$$

The half-width of the concentration gradient is seen to be at a distance $r = 2R$, and the time during which the rate constant is changing most rapidly must be approximately the time required for a B molecule to diffuse this same distance. With $b = 2R$, this time interval is

$$t = \frac{R^2}{(D_A + D_B)}$$

For typical values of $R = 3 \times 10^{-8}$ cm, $D_A + D_B = 5 \times 10^{-5}$ cm^2 sec^{-1}, t is about 2×10^{-11} sec. The assumption made in the development of rate constants for diffusion-limited reactions (page 156)—that a steady-state concentration gradient of B has been established around the A molecules—will be valid as long as the lifetime of A is much greater than 2×10^{-11} sec. The half-life of A in a pseudo-first-order reaction is given by $0.693/k(B)$. If k is 10^{10} M^{-1} sec^{-1}, the steady-state assumption will be valid for concentrations of B below 0.1 M. Rate constants of reactions occurring in times comparable to $R^2/(D_A + D_B)$ are spoken of as "time dependent," for the rate will be abnormally high until the concentration gradient is established (1).

An example of time dependence is provided by studies of the fluorescence of a molecule in an electronically excited state, having a lifetime of about 10^{-9} sec. The excited molecule may either emit light (fluorescence) or react with other molecules and lose its excitation energy (quenching). The competition between these two paths provides a method for investigating the diffusion process, and effects due to the time dependence of rate constants have been observed.

6.5. The Effect of Ionic Strength on Reactions Between Ions

Reaction rate constants for reactions between ions vary with ionic strength in a manner resembling the dependence of equilibrium constants on ionic strength. The effect can be understood in terms of Eq. (6.24) when an approximate form of the electrostatic potential is introduced. The simplest form of the potential—that developed by Debye and Hückel—will do for our purpose (2).

The potential must satisfy Poisson's equation, which is, for a spherically symmetrical charge distribution,

$$r^{-2}\frac{d}{dr}\left[r^2\frac{dE(r)}{dr}\right] = -\frac{4\pi\rho}{\mathscr{D}} \tag{6.43}$$

where $\mathscr{D}$ is the dielectric constant and ρ is the charge density at radius r. The charge density is positive if positive charges predominate, negative if negative charges predominate, and zero if the charges balance. Equation (6.43) is a mathematical description of the fact that lines of force can originate only on charges. For instance, consider an isolated ion A with charge z_A. In the space surrounding the ion, ρ is zero and Eq. (6.43) may be integrated immediately to give

$$\frac{dE(r)}{dr} = \frac{\text{constant}}{r^2}$$

which will be recognized as Coulomb's law,

$$\frac{dE(r)}{dr} = -\frac{z_A e}{\mathcal{D} r^2} \tag{6.44}$$

When other ions are present, ρ will not be zero in the space surrounding A because there will be a tendency for ions of opposite sign to concentrate near A. We will make the assumption that all ions other than A are present as a structureless continuum around A, which allows us to preserve spherical symmetry. The concentration of any type of ion at radius r from A can be determined from Boltzmann's law:

$$n_i = n_i^0 e^{-V(r)/kT}$$

or

$$n_i = n_i^0 \exp\left[-\frac{z_i e E(r)}{kT}\right]$$

where n_i is the number of ions per cm³ at radius r, n_i^0 is the concentration at large r where $V(r)$ is zero, and z_i is the charge on the ion. The charge density is given by the sum of $n_i z_i e$ over all ions present (positive and negative). If there are s species of ions,

$$\rho = \sum_{i=1}^{s} n_i z_i e$$

$$= \sum_{i=1}^{s} n_i^0 z_i e \exp\left[-\frac{z_i e E(r)}{kT}\right]$$

and with Eq. (6.43)

$$r^{-2} \frac{d}{dr}\left[r^2 \frac{dE(r)}{dr}\right] = -\frac{4\pi}{\mathcal{D}} \sum_{i=1}^{s} n_i^0 z_i e \exp\left[-\frac{z_i e E(r)}{kT}\right] \tag{6.45}$$

Here an approximation is needed to facilitate the integration. Debye and Hückel made use of the expansion

$$e^{-x} \cong 1 - x$$

which is valid for small values of x; hence

$$r^{-2} \frac{d}{dr}\left[r^2 \frac{dE(r)}{dr}\right] = -\frac{4\pi}{\mathcal{D}}\left[\sum_{i=1}^{s} n_i^0 z_i e - \sum_{i=1}^{s} \frac{n_i z_i^2 e^2 E(r)}{kT}\right] \tag{6.46}$$

The first sum is zero because the total number of positive and negative charges must balance, so that

$$r^{-2} \frac{d}{dr}\left[r^2 \frac{dE(r)}{dr}\right] = \kappa^2 E(r) \tag{6.47}$$

where

$$\kappa^2 = \frac{4e^2}{kT} \sum_{i=1}^{s} n_i^0 z_i^2 \tag{6.48}$$

The constant κ^{-1} has the dimensions of length and is sometimes referred to as the radius of the ionic atmosphere. The ionic strength of the solution is defined as

$$I = \frac{1}{2} \sum_{i=1}^{s} c_i z_i^2$$

where c_i is the molal concentration of ion i. Consequently,

$$\sum_{i=1}^{s} n_i^0 z_i^2 = \frac{2N_0 d}{1000} I$$

(N_0 is Avogadro's number and d is the density of the solution.) Equation (6.48) can be written

$$\kappa^2 = \frac{8\pi e^2 N_0 d}{1000 \mathscr{D} kT} I \tag{6.49}$$

In water, at 25 °C,

$$\kappa^2 = 1.08 \times 10^{15} I \, cm^{-2} \tag{6.50}$$

The general solution of Eq. (6.47) is

$$E(r) = \frac{C_1 e^{-\kappa r}}{r} + \frac{C_2 e^{\kappa r}}{r}$$

The constant C_2 is zero because $E(r)$ approaches zero at large radius. If A is considered to be a point charge, C_1 may be evaluated by noting that $E(r)$ must approach the value computed for an isolated charge as r approaches zero. Thus from Eq. (6.44)

$$\lim_{r \to 0} \left(\frac{C_1 e^{-\kappa r}}{r}\right) = \frac{z_A e}{\mathscr{D} r} \tag{6.51}$$

from which

$$C_1 = \frac{z_A e}{\mathscr{D}}$$

and the potential is

$$E(r) = \frac{z_A e}{\mathscr{D}} \frac{e^{-\kappa r}}{r} \tag{6.52}$$

Equation (6.52) is valid only for small κ, so nothing is lost by expanding the exponential to give

$$E(r) = \frac{z_A e}{\mathscr{D} r} - \frac{z_A e}{\mathscr{D}} \kappa \tag{6.53}$$

Thus we have an approximate form of the potential describing an ion and its ionic atmosphere, as derived by Debye and Hückel. The first term on the right-hand side of Eq. (6.53) is simply the potential of the isolated ion. The ion atmosphere is represented as a depression of the potential by a constant amount, independent of radius. Of course, this cannot be correct at large radius, since $E(\infty)$ must be zero. Fortunately, however, we are only concerned with the immediate neighborhood of A.

The potential energy of an ion B with charge z_B in the field of A is

$$V(r) = z_B e E(r) = \frac{z_A z_B e^2}{\mathscr{D} r} - \frac{z_A z_B e^2}{\mathscr{D}} \kappa$$

or

$$\begin{aligned} \frac{V(r)}{kT} &= \frac{z_A z_B r_0}{r} - z_A z_B r_0 \kappa \\ &= \frac{V^0(r)}{kT} - z_A z_B r_0 \kappa \end{aligned} \tag{6.54}$$

where r_0 is defined in Eq. (6.32) and $V^0(r)$ is the potential energy of B in the field of an isolated A ion. From Eq. (6.21) we have

$$\beta^{-1} = \exp\left(-z_A z_B r_0 \kappa\right) \int_R^\infty \exp\left(\frac{z_A z_B r_0}{r}\right) r^{-2} \, dr$$

The integral in this expression is just the value of β^{-1} found for ionic reactions at infinite dilution (Eq. 6.31). Denoting this value of β and values of the rate constants at infinite dilution with a superscript ^{0}, we have

$$\beta = \beta^0 \exp\left(z_A z_B r_0 \kappa\right) \tag{6.55}$$

Then, from Eq. (6.25), the diffusion-limited rate constant is

$$k_D = k_D^0 \exp\left(z_A z_B r_0 \kappa\right)$$

With the expression for $V(r)/kT$ given by (6.54), we have

$$k_R e^{-V(r)/kT} = k_R e^{-V^0(r)/kT} \exp\left(z_A z_B r_0 \kappa\right)$$

These rate constants can now be substituted into the rate expression Eq. (6.24) to give

$$k = k^0 \exp\left(z_A z_B r_0 \kappa\right) \tag{6.56}$$

At 25 °C in water, this may be written

$$\log k = \log k^0 + 1.02 \, z_A z_B I^{1/2} \tag{6.57}$$

A somewhat more precise form of Eq. (6.57) for ions of finite size is*

$$\log k = \log k^0 + 1.02 z_A z_B \frac{I^{1/2}}{1 + I^{1/2}} \qquad (6.58)$$

The effect of ionic strength on rate constants is seen to depend on both the magnitude of the charges on the reactants and the signs of the charges. Rate constants will increase with ionic strength for reactions between ions of the same sign, decrease with ionic strength for reactions between ions of opposite sign, and be unaffected by ionic strength for reactions in which one species is uncharged. The variation of some rate constants with the ionic strength function $I^{1/2}/(1 + I^{1/2})$ is given in Fig. 6.2. The lines drawn in the figure are the theoretical slopes for the different reaction pairs as given by Eq. (6.58). Most of the reactions in the figure are much slower than the diffusion limit; however, one diffusion-limited reaction, the reaction between e_{aq} and NO_2^-, is included (see Table 6.2). Three features are apparent from Fig. 6.2. (1) The agreement with the theory is quite good, particularly at low ionic strength. (2) There is some specific effect of sulfate ion on the hydroxide ion—$Co(NH_3)_5Br^{2+}$ reaction. (3) The ionic strength effect on the inversion of sucrose is not quite nil, as was expected from Eq. (6.58).

The specific effect of sulfate ions probably results from formation of an ion-pair complex, $Co(NH_3)_5Br^{2+}SO_4^{2-}$, which reacts less readily with hydroxide ion because it is uncharged (3). The formation of this complex is highly favored, for both ions are doubly charged. Such specific ion effects are a constant danger in ionic strength experiments. Whenever possible the experimenter should avoid using multiply charged ions to vary the ionic strength and should avoid using ionic strengths above 0.05 M. At higher concentrations, ionic strength effects may disappear altogether or may even be reversed.

The effect of ionic strength on the sucrose inversion is an order of magnitude smaller than would be found if sucrose had a negative charge; still, it is real and is due to the large dipole moment of sucrose. Dipolar interactions were not considered in the derivation of the Debye-Hückel equation.

Within the approximation involved in Eq. (6.58), the ionic strength effect is the same for diffusion-limited reactions as for slower reactions but with one equalification. Diffusion-limited reactions are studied by

---

*The constant of integration C_1 in Eq. (6.51) should be evaluated at the surface of A, in which case it is (2)

$$C_1 = \frac{z_A e}{\mathscr{D}} \frac{e^{a\kappa}}{1 + a\kappa}$$

If a, the radius of A, is 3 Å, $a\kappa = I^{1/2}$.

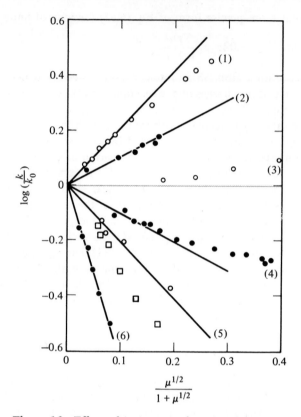

Figure 6.2. *Effect of ionic strength on rate constants.*
(1) $BrCH_2COO^- + S_2O_3^{2-}$; *(2)* $e_{aq} + NO_2^-$; *(3)*
$H_3O^+ + C_{12}H_{22}O_{11}$ *(inversion of sucrose);* *(4)*
$H_3O^+ + Br^- + H_2O_2$; *(5)* $OH^- + Co(NH_3)_5Br^{2+}$
[ionic strength varied with $NaBr$ *(circles) and* Na_2SO_4
(squares)]; *(6)* $Fe(H_2O)_6^{2+} + Co(C_2O_4)_3^{3-}$. *Data
from (1) V. K. LaMer and R. W. Fessenden, J.*
Am. Chem. Soc. **54**. *2351 (1932); (2)G. Czapski and*
H. A. Schwarz, J. Phys. Chem. **66**, *471 (1962); (3) C. F.*
Kautz and A. L. Robinson, J. Am. Chem. Soc. **50**,
1022 (1928); (4) R. S. Livingston, J. Am. Chem. Soc.
48, *53 (1926); (5) A. R. Olson and T. R. Simonson,*
J. Chem. Phys. **17**, *1167 (1949); (6) J. Barnett and*
J. H. Baxendale, Trans. Faraday Soc **52**, *210 (1956).*

generating one of the reactants suddenly in solution; when formed, it
will not be surrounded by the normal ionic atmosphere. The time
required to develop the atmosphere will be approximately that required

for an ion to diffuse a distance equal to the radius of the ionic atmosphere, κ^{-1} (neglecting motion by conduction). By Eq. (6.41), this time is

$$t = \frac{1}{4D\kappa^2}$$

or

$$t \cong \frac{10^{-11}}{I} \text{ sec}$$

This time is comparable to the half-lifetime of the reaction $0.69/k_D c$ if the only ions present are the reactants. For this reason, an inert salt is added in ionic-strength effect studies, so that I is much greater than the concentration of reactants.

6.6. *Slow Reactions*

The majority of reactions that have been studied in solution are considerably slower than the diffusion limit for the reaction.* Slow reactions are characterized by the inequality

$$k_R e^{-V(R)/kT} \ll k_D$$

in which case Eq. (6.24) reduces to

$$k = k_R e^{-V(R)/kT} \tag{6.59}$$

The generation of Eq. (6.59) from Eq. (6.14), through the development of the effect of diffusion on reaction rates, does not constitute a derivation, however. The result was built into the development at the beginning, when the ionic mobility was related to the diffusion coefficient by means of Boltzmann's law. Equation (6.59) follows immediately upon substituting Eq. (6.11) in Eq. (6.23) [$C_i = (B)_R$ and $C_i^0 = (B)$]. In this way, it is seen that Eq. (6.59) is more general than is implied by the derivation from Eq. (6.24), since no assumption of spherical symmetry need be made.

Equation (6.59) is most useful in discussing the reactions between ions. The potential energy is given by Eq. (6.30), which with Eq. (6.59) gives

$$k = k_R \exp\left(\frac{-z_A z_B e^2}{\mathscr{D}kTR}\right) \tag{6.60}$$

*Note that the study of slow reactions had a head start. The investigation of rapid reactions waited many years for the development of techniques with short-time resolution.

We cannot make an a priori estimate of k_R for any actual reaction, but Eq. (6.60) does predict the dependence of k on the dielectric constant of the medium if it is assumed that k_R is independent of the properties of the solvent.

The dielectric constant can be adjusted by changing the solvent composition. For example, the dielectric constant of a mixture of methanol and water can be varied continuously from 78 in pure water to 32 in pure methanol. The logarithm of the rate constant of an ionic reaction should vary inversely with the dielectric constant, and indeed it often does over a limited range of composition. The slope of the line should yield a value for R that is of the magnitude of atomic dimensions. Some typical data are given in Table 6.3.

Table 6.3. *Reaction Radii for Slow Ionic Reactions*

Reaction	Solvents	R, Å
Tetrabromophenolpthalein + OH^-[a]	H_2O + EtOH	1.2
	H_2O + MeOH	1.5
$BrCH_3COO^- + S_2O_3^{2-}$[a]	H_2O[b]	5.1
$(CH_3)_3S^+ + OH^-$[c]	H_2O + EtOH	1.4

[a]E. S. Amis, *Solvent Effects on Reaction Rates and Mechanisms*, Academic Press, New York 1966, p. 7.
[b]The dielectric constant was varied by adding glycine, sucrose, or urea.
[c]From data of J. L. Gleave, E. D. Hughes, and C. K. Ingold, *J. Chem. Soc.*, 236 (1935).

Equation (6.60) is only valid over a limited range of solvent composition, because its derivation neglected all but electrostatic interactions and the molecular nature of the solvent. Both of these are oversimplifications of the actual ion-solvent interaction. For example, the proton is a much different species in water than it is in methanol. In aqueous solution, the proton is usually formulated as H_3O^+ to indicate intimate association with at least one solvent molecule, whereas in methanol it forms $CH_3OH_2^+$. The free energy difference between H_3O^+ and $CH_3OH_2^+$ cannot be accounted for by classical considerations like Coulomb's law. Similar, though weaker, associations take place between other ions and solvents.

The effect of dielectric constant on reactions of ions with polar molecules can also be predicted by using Eq. (6.59). The most stable configuration for a dipole in the field of an ion is with the negative end pointing toward a positive ion or the positive end pointing toward a negative ion. In this position, the potential energy of a molecule with dipole moment μ at a distance r from an ion of charge ze is

$$V(r) = \frac{-ze\mu}{\mathscr{D}r^2}$$

which in Eq. (6.59) gives

$$k = k_R \exp\left(\frac{ze\mu}{\mathscr{D}R^2kT}\right) \tag{6.61}$$

The ratio of rate constants in two solvents of different dielectric constants $\mathscr{D}_1$ and $\mathscr{D}_2$ is

$$\frac{k_1}{k_2} = \exp\left(\frac{ze\mu}{R^2kT}\right)\left(\frac{1}{\mathscr{D}_1} - \frac{1}{\mathscr{D}_2}\right) \tag{6.62}$$

so the reaction will be faster in solvents of lower-dielectric constant. If the dipole must approach the ion in an unfavorable position—that is, positive end of the dipole toward a positive ion—the sign of the exponent in Eq. (6.62) will be reversed and the reaction will be slower in solvents of lower-dielectric constant. A typical reaction might involve a molecule with a dipole moment of 3 debyes (3×10^{-18} esu cm) and a van der Waals separation from the positive ion, R, of 3×10^{-8} cm. For such a reaction, the expected rate constant ratio in ethanol ($\mathscr{D} = 25$) to water ($\mathscr{D} = 78$) is only 3. A ratio of this magnitude is observed experimentally, for example, in the acid-catalyzed inversion of sucrose (4). The effect is very small compared to the effect of solvents on reactions between ions. Equation (6.61) is similar to Eq. (6.60) except that the charge on one of the ions, z_B, has been replaced by μ/R, which is only about one-fifth as large.

Equation (6.61) appears to be valid in a very limited number of reactions. There are many reasons for this situation. In approaching the ion, the polar molecule usually displaces a polar solvent molecule from the solvation sphere of the ion; the difference in free energies between the solvent-ion complex and the reactant-ion complex should probably be used in predicting the rate constant, as suggested by Benson (5). Moreover, specific effects of the solvent on the ion, outside the framework of classical electrostatics, are likely to be more profound than the ion-dipole interaction. There is no quantitatively satisfactory, simple theory for ion-dipole reactions.

Reactions between neutral species (except those resulting in ionization) do not exhibit any general solvent effect. The total electrostatic interaction between two dipoles is small; consequently, the dependence on the solvent is also small. There are many specific solvent effects on reactions between neutral species, however, just as on reactions of ions. This point is clearly shown by the data in Table 6.4 on the relative rates at which chlorine atoms attack the tertiary and primary hydrogens of 2, 3–dimethyl butane. The chlorine atoms are formed by the photolysis

Table 6.4. *Solvent Effects in the Photochlorination of 2, 3–dimethyl butane at 55 °C*[a]

Additive	$k_{6.63}/k_{6.64}$	Additive	$k_{6.63}/k_{6.64}$
None	.61	Benzene	2.4
Carbon tetrachloride	.58	Nitrobenzene	.8
Nitromethane	.55	Methylbenzoate	1.7
n-Butyric acid	.68	Chlorobenzene	1.7
Propionitrile	.67	Trifluoromethylbenzene	1.2
Dimethyl sulfolane	.60	t-Butylbenzene	4.0
		Toluene	2.6

[a]G. A. Russell, *J. Am. Chem. Soc.* **80**, 4987 (1958).

of chlorine and react according to

$$Cl + C_6H_{14} \longrightarrow H_3C-\overset{\overset{\displaystyle H_3C}{|}}{\underset{\underset{\displaystyle H}{|}}{C}}-\overset{\overset{\displaystyle CH_3}{|}}{C}-CH_3 + HCl \qquad (6.63)$$

$$Cl + C_6H_{14} \longrightarrow H_3C-\overset{\overset{\displaystyle H_3C}{|}}{\underset{\underset{\displaystyle H}{|}}{C}}-\overset{\overset{\displaystyle CH_3}{|}}{\underset{\underset{\displaystyle H}{|}}{C}}-CH_2 + HCl \qquad (6.64)$$

The hydrocarbon radicals then react with chlorine to form the respective tertiary and primary chlorides and chlorine atoms, so that the overall reaction is a chain reaction. The ratio $k_{6.63}/k_{6.64}$ is equal to the product ratio of tertiary chloride to primary chloride, independent of the kinetics of the initiation and termination steps of the chain. The rate constant ratio was studied with various additives at 4 M concentration (about 50 percent by weight). The additives in the first column had no appreciable effect on the ratio, even though the dielectric constants of the additives range up to 39 for nitromethane. This is in agreement with the conclusions that the solvent affects neither rate very much. The additives in the second column have a large effect on the ratio, however. Obviously at least one of the two rate constants is solvent dependent. It is most probable that both decrease with added solute, reaction (6.64) being affected the most. Note that the additives in the second column are aromatic. Apparently aromatic compounds form complexes with chlorine atoms; and the more stable the complex, the greater the selectivity of abstraction in favor of tertiary hydrogens.

Another important conclusion from Eq. (6.59) is that the rates of slow reactions in solution are independent of viscosity. This independence is seen in reaction (6.65)

$$I^{131} + ICH\!=\!CHI \longrightarrow I^{131}CH\!=\!CHI + I \tag{6.65}$$

The rate of this exchange reaction can be measured when I_2^{131} is photo-lyzed in the presence of vinyl iodide (6). The rate constant for reaction (6.65) is $5.7 \times 10^3\ M^{-1}\ \text{sec}^{-1}$ when measured in n-hexane, 5.2×10^3 $M^{-1}\ \text{sec}^{-1}$ in carbon tetrachloride, and $6.5 \times 10^3\ M^{-1}\ \text{sec}^{-1}$ in hexachlorobutadiene. The rate is independent of the nature of the solvent, as expected, even though the viscosity of hexachlorobutadiene is 100 times that of n-hexane.

6.7. *Activated Complex Theory Formulation*

So far in the discussion of reactions in liquids, we have considered only the effect of the solvent on the reactants as they form a reacting pair in contact. Nothing has been said about the transition state of the reaction. Rate constants in liquids can also be viewed from the standpoint of activated complex theory. This approach is not exactly an alternate presentation to the development in Sections 6.1 to 6.6, for certain effects, such as diffusion-limited reactions, cannot be adequately dis-cussed in terms of activated complex theory, nor can other effects, such as comparison of gas- and liquid-phase rate constants be satisfactorily discussed in terms of the earlier physical model.

Consider the reaction

$$A + B \longrightarrow (AB)^{\ddagger} \longrightarrow \text{products}$$

The equilibrium constant for formation of the activated complex of this reaction should be expressed by the activities of the species instead of the concentrations. Thus

$$K^{\ddagger} = \frac{a^{\ddagger}}{a_A a_B} \tag{6.66}$$

where $a^{\ddagger}$ is the activity of the activated complex and a_A and a_B are the activities of the reactants. The rate of the reaction, according to acti-vated complex theory, is proportional to the *concentration* of the acti-vated complex:

$$\mathscr{R} = \frac{kT}{h}(AB^{\ddagger}) \tag{6.67}$$

The concentration of the activated complex may be found from Eq. (6.66) by introducing the activity coefficients γ_A, γ_B, and $\gamma^{\ddagger}$,

$$(AB^{\ddagger}) = K^{\ddagger}\frac{\gamma_A \gamma_B}{\gamma^{\ddagger}}(A)(B)$$

and from Eq. (6.67)

$$\mathscr{R} = \frac{kT}{h} K^{\ddagger} \frac{\gamma_A \gamma_B}{\gamma^{\ddagger}} (A)(B)$$

or

$$k = \frac{kT}{h} K^{\ddagger} \frac{\gamma_A \gamma_B}{\gamma^{\ddagger}} \qquad (6.68)$$

An a priori estimate of $K^{\ddagger}$ is not needed in order to compare rates in different solvents or phases, since $(kT/h)K^{\ddagger}$, which may be defined as k_0, the rate constant in some arbitrary reference state, can be determined by experiment. Introducing k_0 into Eq. (6.68) gives

$$k = k_0 \frac{\gamma_A \gamma_B}{\gamma^{\ddagger}} \qquad (6.69)$$

The activity coefficients are defined as unity in the reference state, which may be chosen to suit the experiment. For example, the dilute gas is a convenient reference state in comparing rates between gas and liquid phase, whereas the hypothetical state of infinite dilution is convenient for ionic reactions.

With this model, the activity coefficient of the activated complex has the same meaning as the activity coefficient of any ordinary species. Note, however, that a particular activated complex is unique to a given reaction. Therefore, unlike activity coefficients of ordinary chemical species, $\gamma^{\ddagger}$ cannot be determined in one reaction and used in another. Nor can it be found by the usual techniques, such as osmotic pressure or potentiometric measurements. In some cases, however, $\gamma^{\ddagger}$ can be predicted by analogy between the activated complex and stable compounds or by the use of thermodynamic theories.

An example of the use of analogy may be found in unimolecular decompositions, such as the dissociation of azo-bis-isobutyronitrile

$$\underset{\overset{|}{R}}{\overset{\overset{R}{|}}{NC\!-\!C\!-\!N}}\!=\!\underset{\overset{|}{R}}{\overset{\overset{R}{|}}{N\!-\!C\!-\!CN}} \longrightarrow N_2 + 2\,\underset{\overset{|}{R}}{\overset{\overset{R}{|}}{NC\!-\!C}}\cdot \qquad (6.70)$$

where R is an isobutyl group. Since the reaction is unimolecular, the proper equation corresponding to Eq. (6.69) is

$$k = k_0 \frac{\gamma_A}{\gamma^{\ddagger}}$$

The activated complex in this reaction will be very similar to the parent compound, with a few bonds slightly elongated; These bonds are not very polar, for the products of the dissociation are free radicals and not ions. The reactant itself is the best possible analog for the activated complex; and even though the nitrile groups interact strongly with various solvents, the interaction should be the same for the reactant and activated complex. The activity coefficient ratio, $\gamma_A/\gamma^{\ddagger}$, and hence the reaction rate constant, should be independent of the nature of the solvent. According to expectations, the rate constant is 1.55×10^{-4} sec^{-1} in toluene, 1.52×10^{-4} sec^{-1} in acetic acid, and 1.98×10^{-4} sec^{-1} in nitrobenzene, all at 80 °C (7).

The reactions of ions can be discussed in terms of activated complex theory as well as in terms of the earlier model (Section 6.6). The activity coefficient of an ion depends on the extent to which the free energy of the ion, G, differs from some assumed ideal behavior, G_{ideal},

$$\Delta G = G - G_{ideal} = kT \ln \gamma \tag{6.71}$$

(if ΔG is computed for a single ion). The part of the free energy of an ion with radius a which may be attributed to its charge, ze, may be found from the work done to charge the ion from 0 to ze. The contribution to G from an element of charge de is

$$dG = de\, E(e, a)$$

From Eq. (6.53)

$$E(e, a) = e\left(\frac{1}{\mathscr{D}a} - \frac{\kappa}{\mathscr{D}}\right)$$

so that

$$G = G' + \int_0^{ze} \left(\frac{1}{\mathscr{D}a} - \frac{\kappa}{\mathscr{D}}\right) e\, de$$

$$= G' + \frac{z^2 e^2}{2\mathscr{D}a} - \frac{z^2 e^2}{2\mathscr{D}}\kappa \tag{6.72}$$

where G' includes all parts of the free energy not due to the interaction of the charge with the medium.

If our reference state is taken as an infinitely dilute solution of ions, κ is zero and

$$G_{ideal} = G' + \frac{z^2 e^2}{2\mathscr{D}a}$$

$$\Delta G = -\frac{z^2 e^2}{2\mathscr{D}}\kappa$$

and from (6.71)

$$\ln \gamma = -\frac{z^2 r_0 \kappa}{2}$$

where r_0 is defined by Eq. (6.32). In water at 25 °C,

$$\log \gamma = -0.51\, z^2 I^{1/2} \tag{6.73}$$

The effect of ionic strength on reactions between ions can be seen by rewriting Eq. (6.69) as

$$\log k = \log k_0 + \log \gamma_A + \log \gamma_B - \log \gamma^\ddagger \tag{6.74}$$

$$= \log k_0 - .51 I^{1/2}[z_A^2 + z_B^2 - (z_A + z_B)^2]$$

$$\log k = \log k_0 + 1.02\, z_A z_B I^{1/2}$$

which is the same as Eq. (6.57).

Next we will reconsider the effect of solvent on reactions between ions at negligibly low ionic strength. Equation (6.72) becomes

$$G = G' + \frac{z^2 e}{2\mathscr{D} a}$$

and taking our reference state as a solvent of infinite dielectric constant ($G_{\text{ideal}} = G'$), we have

$$\Delta G = \frac{z^2 e^2}{2\mathscr{D} a} \tag{6.75}$$

The activity coefficient is given by

$$\ln \gamma = \frac{z^2 e^2}{2\mathscr{D} kT a}$$

and from (6.74)

$$\ln k = \ln k_0 + \frac{e^2}{2\mathscr{D} kT}\left[\frac{z_A^2}{a_A} + \frac{z_B^2}{a_B} - \frac{(z_A + z_B)^2}{a^\ddagger}\right] \tag{6.76}$$

Equation (6.76) is similar to, but not identical with, Eq. (6.60), which may be written

$$\ln k = \ln k_0 - \frac{e^2}{\mathscr{D} kT}\frac{z_A z_B}{R} \tag{6.77}$$

The constant k_0 is easily identified with k_R, but the term in square brackets is not the same in the two equations. The derivation leading to Eq. (6.60) was based on a model of two charged spheres that react when their centers are separated by a distance R, whereas the model leading to Eq. (6.76) involves a spherical activated complex. Both equations predict the same form for the dependence of rate constant on dielectric constant. This is the significant experimental effect.

Both the $\ln k_R$ term and the electrostatic term in Eq. (6.77) [or Eq. (6.66)] will contribute to the temperature dependence of k. The temperature dependence of the electrostatic term will be determined by the product $\mathscr{D} T$, and it is interesting that for solvents usually used in the study of ionic reactions (i.e., water, alcohols, nitrobenzene, etc.) the product $\mathscr{D} T$ actually decreases with increasing temperature. In water, $\mathscr{D} T$ is 2.404×10^4 at 0 °C and 2.342×10^4 at 25 °C. The part of the free energy of activation due to electrostatic interaction, $\Delta G_{el}^{\ddagger}$, is largely manifested as an entropy of activation, $\Delta S_{el}^{\ddagger}$. The large dielectric constants of the liquids are due to an ordering of the solvent molecules in the electric field. If the reacting ions are oppositely charged, the order will be greatly reduced in the activated complex; therefore $\Delta S_{el}^{\ddagger}$ is positive. Conversely, if the ions have the same sign, the charges reinforce one another and the solvent will be more ordered, or $\Delta S_{el}^{\ddagger}$ will be negative. From Eq. (6.77)

$$\Delta G_{el}^{\ddagger} = \frac{e^2 z_A z_B}{\mathscr{D} R}$$

and

$$\Delta S_{el}^{\ddagger} = -\left(\frac{\partial \Delta G_{el}^{\ddagger}}{\partial T}\right)_P = -\frac{e^2 z_A z_B}{\mathscr{D}^2 R}\frac{\partial \mathscr{D}}{\partial T}$$

In water, $\partial \mathscr{D}/\partial T$ is -0.37 at 25 °C, so if we assume that R is 2×10^{-8} cm, $\Delta S_{el}^{\ddagger}$ is about 7×10^{-16} erg per °C per ion for unit charges, or $+10$ eu for oppositely charged ions, -10 eu for similarly charged ions.

Ionization reactions are the reverse of ion recombination and (in Section 6.6) were specified as exceptions to the generalization concerning reactions of neutral molecules. The solvent effect is on the activated complex of the reaction, which may be considered as a very strong dipole. Activated complex theory can be used to predict in a qualitative fashion (and perhaps semiquantitatively) that the rate of ionization will be facilitated in solvents of high dielectric constant. Other, much more specific, effects are also operating, however. Among these are diffusion effects (many of the reverse reactions are diffusion limited). At present, solvent effects on ionization reactions are best handled by empirical correlations (as in Section 7.6).

6.8. Comparison of Gas-Phase and Liquid-Phase Rate Constants

Perhaps the most drastic change in solvent would be from the gas phase, where there is no solvent, to the liquid phase. Some generalizations about such a change can be made with the help of Eq. (6.69). The gas phase is a convenient reference state for the comparison.

The activity coefficient of the gas at pressures up to several atmospheres can be taken as unity for our purpose. For a solution in equilibrium with a gas, the activity of the gas is the same in both phases, because the standard state is the same for both phases. Consequently,

$$\gamma_{sol} C_{sol} = C_{gas}$$

or

$$\gamma_{sol} = \frac{C_{gas}}{C_{sol}} \tag{6.78}$$

The activity coefficient γ_{sol} is independent of pressure if the mole fraction of the gas in solution is small (γ_{sol}^{-1} is called the Ostwald absorption coefficient). Values of C_{gas}/C_{sol} for hydrogen, oxygen, and carbon monoxide are given in Table 6.5. The concentration ratios, and con-

Table 6.5. *Ratio of Concentration in Gas Phase to Concentration in Solution, 25°C*

Solvent	Gas		
	H_2	O_2	CO
H_2O	52	30	43
CH_3OH	11	5	5
C_6H_6	13	6	6

sequently the activity coefficients, for these gases are considerably greater than unity. This is not true for gases like hydrogen chloride, which react with the solvent, or for gases with relatively high boiling points, such as ammonia or carbon dioxide. We shall restrict this discussion to gases with a low affinity for the solvent.

If all the activity coefficients in Eq. (6.69) were of the magnitude shown in Table 6.5, the rate constant would be 10 to 30 times larger in solution than in the gas. More generally, if the interaction of the activated complex with the solvent is similar to the interaction of one of the reactants with the solvent, their activity coefficients will cancel and the ratio of the liquid phase to gas phase rate constants will be given approximately by the activity coefficient of the remaining reactant. In unimolecular reactions, almost complete cancellation of activity coefficients is to be expected.

Few rate constants for bimolecular reactions are known in both the liquid and gas phases, but three examples are given in Table 6.6. The reactants and activated complex in the reaction of hydrogen atoms with deuterium probably fall in the category of gases given in Table 6.5, and a factor of about 30 in rate is to be expected. In the second reaction, hydrogen bonding will result in strong interactions between the solvent, water, and both the hydroxyl radical and the activated complex, H_2OH. To a first approximation, activity coefficients of these two species will cancel, leaving the activity coefficient of hydrogen (52) as the predicted ratio of rate constants. The transition state in the third reaction is COOH, and it is not obvious that $\gamma^{\ddagger}$ will equal γ_{OH} in aqueous solution. Again,

Table 6.6. *Comparison of Gas-Phase and Liquid-Phase Reaction Rate Constants (in units of $M^{-1} sec^{-1}$), at 25°C*

Reaction	k_{H_2O}[a]	k_{gas}[b]	k_{H_2O}/k_{gas}
$H + D_2 \rightarrow HD + D$	8×10^5	2×10^4	40
$OH + H_2 \rightarrow H_2O + H$	5×10^7	3×10^6	17
$OH + CO \rightarrow COOH$	6×10^8	1×10^8	6

[a]Reviewed by M. Anbar and P. Neta, *Int. J. Applied Radiation and Isotopes* **61**, 227 (1965).
[b]F. Kaufman, *Ann. Rev. Phys. Chem.* **20**, 45 (1969).

both are capable of hydrogen bonding with water, so that it is not unreasonable that the rate constant ratio is of the same magnitude as that for the other two reactions.

From thermodynamics, the activity coefficient in Eq. (6.78) is related to the free energy of solution of the gas according to

$$\gamma_{sol} = \exp\left(\frac{-\Delta G_{sol}}{RT}\right) = \exp\left(\frac{-\Delta H_{sol}}{RT}\right)\exp\left(\frac{\Delta S_{sol}}{R}\right)$$

For most gases, ΔH_{sol} is negative; for the examples in Table 6.5 it is small. The entropy of solution is related to the molar volume (V^0) and a quantity called the free volume (V_f) by the expression

$$\exp\left(\frac{\Delta S_{sol}}{R}\right) = \frac{V^0}{V_f}$$

The free volume of the solute is defined as the volume per mole less the actual volume of the molecules themselves. This quantity is of the order of one percent of the molar volume, and this ratio is the source of the large values of γ_{sol} shown in Table 6.5.

6.9. The Effect of Pressure on Rate Constants

As mentioned in Chapter 1, rate constants do vary with pressure. There is nothing about the theory of the effect that limits it to reactions in solution, but the pressures involved (commonly several thousand atmospheres) create experimental problems for work with gases. Consequently, only the liquid phase is studied in practice.

The thermodynamic formulation of absolute rate theory may be written [Eq. (4.51)]

$$\ln k = \ln \frac{kT}{h} - \frac{\Delta G^{\ddagger}}{RT}$$

(Activity coefficients, etc., are absorbed in $\Delta G^{\ddagger}$.) The effect of pressure on k may be found by taking the partial derivative of $\ln k$ with respect to pressure at constant temperature.

$$\left(\frac{\partial \ln k}{\partial P}\right)_T = -\frac{1}{RT}\left(\frac{\partial \Delta G^{\ddagger}}{\partial P}\right)_T$$

Since

$$\left(\frac{\partial \Delta G}{\partial P}\right)_T = \Delta V$$

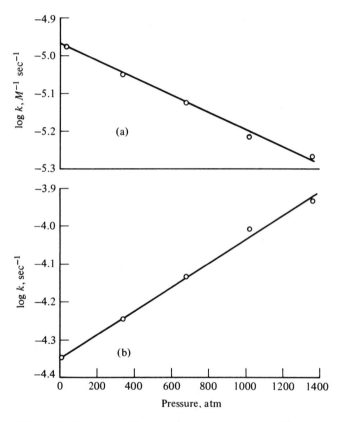

Figure 6.3. *Pressure effect on rate constants in solution. I,*
$(t\text{-}C_5H_{11})(CH_3)_3N^+I^- + OH^- \longrightarrow (CH_3)_3N + H_2O +$
methylbutene, in ethanol at 85°C, $\Delta V^{\ddagger} = 15$ cm³/mole. II,
$t\text{-}C_5H_{11}Cl + C_2H_5OH \longrightarrow t\text{-}C_5H_{11}OH + C_2H_5Cl,$ *in*
80% aqueous ethanol at 34°C, $\Delta V^{\ddagger} = -18$ cm³/mole. [Data
from K. R. Brower and J. S. Chen, J. Am. Chem. Soc. 87,
3396 (1965).]

we have

$$\left(\frac{\partial \ln k}{\partial P}\right)_T = \frac{-\Delta V^{\ddagger}}{RT} \tag{6.79}$$

where $\Delta V^{\ddagger}$ is the difference in molal volume between the reactants and the activated complex:

$$\Delta V^{\ddagger} = V^{\ddagger} - \sum V_{\text{reactants}}$$

The dependence of $\ln k$ on pressure for two reactions is given in Fig. 6.3. The linear dependence of $\ln k$ on pressure is good in these examples, but often there is a pronounced curvature and an initial slope must be estimated.

The volume of activation can be a useful quantity in determining reaction mechanism. Consider, for instance, a unimolecular reaction proceeding by the breaking of a bond. The bond can be represented as a cylinder with a volume $\pi R^2 l$, where R is a van der Waal's radius of about 2 Å and l is the bond length. In the activated complex, the bond is stretched by about one angstrom, so that the volume increase upon activation would be about 12 Å^3/molecule or 7 cm^3/mole. Reactions (6.80) through (6.82) in Table 6.7 are bondcleavage reactions.

Table 6.7. *Volumes of Activation*[a]

	Reaction	Solvent	$\Delta V^{\ddagger}$ cm^3 mole^{-1}
(6.80)	$(CH_3)_3COOC(CH_3)_3 \rightarrow$ $2(CH_3)_3CO$	Cyclohexane	+6.7
(6.81)	$(CH_3)_2C-N=N-C(CH_3)_2 \rightarrow$ $2(CH_3)CN + N_2$	Toluene	+3.8
(6.82)	$C_6H_5COOOOCC_6H_5 \rightarrow$ $2C_6H_5COO$	Carbon tetrachloride	+11.
(6.83)	$2C_5H_6(\text{cyclopentadiene}) \rightarrow$ $C_{10}H_{12}$	n-butyl chloride	−22.
(6.84)	$OH^- + (C_2H_5)_3S^+ \rightarrow$ $C_2H_5OH + (C_2H_5)_2S$	Water	+10.
(6.85)	$OH^- + N(C_2H_5)_4^+ \rightarrow$ $CH_3OC_2H_5 + N(C_2H_5)_3$	Methanol	+20.

[a]From Reference 8.

Volume changes of the solvent due to the reaction are also included in $\Delta V^{\ddagger}$. In the process of ionization, there is a charge separation in the activated complex that is not present in the parent molecule. The presence of a charge induces a contraction of the solvent, called electrostriction, so that the volume of activation should be negative (8). Typical magnitudes are -10 to -20 cm^3/mole.

Volume changes for the reverse processes may be found by noting that the same activated complex is reached as in the forward reaction. By conservation of volume,

$$\Delta V^{\ddagger}_{\text{forward}} - \Delta V^{\ddagger}_{\text{reverse}} = \sum V_{\text{products}} - \sum V_{\text{reactants}}$$

since the activation volumes for the forward and reverse reactions are measured in opposite directions. The molar volumes of the products and reactants are generally known from density measurements. Typically, $\Delta V^{\ddagger}$ for bond formation is about $-10 \text{ cm}^3 \text{ mole}^{-1}$ [see reaction (6.83) in Table 6.7], and for neutralization, about $20 \text{ cm}^3 \text{ mole}^{-1}$ [reactions (6.84) and (6.85)].

References

1. R. M. Noyes, in *Progress in Reaction Kinetics*, G. Porter (Ed.), Pergamon Press, Oxford, 1961, vol. 1, p. 129.
2. See *The Physical Chemistry of Electrolytic Solutions*, 3rd ed., by H. S. Harned and B. B. Owen, Van Nostrand Reinhold, New York, 1958.
3. C. W. Davies, in *Progress in Reaction Kinetics* G. Porter (Ed.), Pergamon Press, Oxford, 1961, vol. 1, p. 161.
4. E. S. Amis and F. C. Holmes, *J. Am. Chem. Soc.* **63**, 2231 (1941).
5. S. W. Benson, *The Foundations of Chemical Kinetics*, McGraw-Hill, New York, 1960, p. 536.
6. H. Rossman and R. M. Noyes, *J. Am. Chem. Soc.* **80**, 2410 (1958).
7. C. Walling, *Free Radicals in Solution*, John Wiley & Sons, New York, 1957, p. 512.
8. W. J. LeNoble, in *Progress in Physical Organic Chemistry*, A. Streitweiser and R. W. Taft (Eds.), Interscience Publishers, New York, 1967 vol. 5, p. 207.

Problems

6.1. Compute the mobilities and diffusion constants for the ammonium and hydroxide ions from the following equivalent conductivities ($\text{cm}^2 \text{ ohm}^{-1} \text{ equiv}^{-1}$):

	13 °C	*20 °C*	*25 °C*
OH^-	158	180	198
NH_4^+	–	66	–

6.2. These rate constants have been measured:

Reaction	Temperature, °C	$k, M^{-1} sec^{-1}$
$OH^- + NH_4^+ \longrightarrow H_2O + NH_3$	20	3.4×10^{10}
$OH^- + \bigcirc_{NO_2}^{OH} \longrightarrow H_2O + \bigcirc_{NO_2}^{O^-}$	13	1.1×10^{10}
$OH^- + HCN \longrightarrow H_2O + CN^-$	25	3.7×10^9

Compare them with diffusion-limited rate constants using either Eq. (6.27) or Eq. (6.33) as appropriate, and also using diffusion constants calculated in problem 6.1 and assumed values of $1.5 \times 10^{-5} cm^2 sec^{-1}$ at 13 °C and $2 \times 10^{-5} cm^2 sec^{-1}$ at 25 °C for orthonitrophenol and hydrogen cyanide, respectively. The van der Waals radii are approximately 1.5 Å for OH^-, 1.5 Å for NH_4^+, 2.5 Å for *o*-nitrophenol, and 1.5 Å for HCN. The dielectric constant of water is 80.4 at 20 °C. (From Reference c, Table 6.1).

6.3. The hydrolyses of ethyl esters

$$RCOOEt + H_2O \longrightarrow RCOOH + EtOH$$

are catalyzed by acid and by base, the rate laws being

$$\mathscr{R} = k_A(H^+)(RCOOEt)$$

and

$$\mathscr{R} = k_B(OH^-)(RCOOEt)$$

The following rate constants have been measured in aqueous solution:

Reaction	Temperature	$k, M^{-1} sec^{-1}$
(a) $H^+ + (C_2H_5)_3N^+CH_2COOC_2H_5$	40 °C	1.4×10^{-7}
(b) $H^+ + CH_3COOC_2H_5$	40 °C	4.2×10^{-4}
(c) $CH^- + (C_2H_5)_3N^+CH_2COOC_2H_5$	25 °C	32
(d) $OH^- + CH_3COOC_2H_5$	25 °C	0.11

Do the rate constants change in the expected manner in going from an uncharged to a charged ester? Assume that, in applying Eq. (6.60), k_R for reaction (a) is given by k for reaction (b) and k_R for reaction (c) is given by k for reaction (d). Calculate reaction radii for reactions (a) and (c). The dielectric constant of water is 78.5 at 25 °C and 73.3 at 40 °C. [R. P. Bell and F. L. Lindars, *J. Chem. Soc.*, 4601 (1954).]

6.4. The rate constant for the Menschutkin reaction

$$C_6H_5N(CH_3)_2 + C_2H_5I \longrightarrow (C_6H_5N(CH_3)_2C_2H_5)^+I^-$$

exhibits the following pressure dependence at 52.5 °C in methanol and nitrobenzene solvents:

| | $k \times 10^5$, M^{-1} sec^{-1} | |
Pressure, atm	CH$_3$OH	C$_6$H$_5$NO$_2$
1	3.18	2.24
1500	12	6.15
2300	–	9.51
2600	–	11.3
2875	27.7	–
4925	57.6	–

Estimate $\Delta V^{\ddagger}$ for the reaction at one atmosphere. (Note that these examples of pressure dependence do not yield a constant $\Delta V^{\ddagger}$ and are perhaps more typical than the data in Fig. 6.3.) [A. P. Harris and K. E. Weale, *J. Chem. Soc.*, 146 (1961).]

6.5. The conductivity of *n*-hexane, induced by x rays, decays when the x rays are turned off because the ions are recombining:

$$R^+ + X^- \longrightarrow \text{products}$$

The specific conductivity, l, follows second-order kinetics:

$$\frac{1}{l} + \frac{1}{l_0} = \frac{kt}{u_{R^+} + u_{X^-}}$$

and $k/(u_{R^+} + u_{X^-})$ is found experimentally to be 0.98×10^{-6} ($\pm 5\%$) volt cm at 24 °C (k in units of molecules^{-1} cm^3 sec^{-1}). Compare this with the computed value, using Eq. (6.13) and the simplified form ($r_0 \gg R$) of Eq. (6.33):

$$k_D = 4\pi(D_A + D_B)r_0$$

The dielectric constant of *n*-hexane is 1.87. How large is the reaction radius? The positive and negative ion mobilities were found to be 0.68×10^{-3} and 1.3×10^{-3} cm^2 volt^{-1} sec^{-1} in a separate experiment. What is k in units of M^{-1} sec^{-1}? [A. Hummel and A. O. Allen, *J. Chem. Phys.* **44**, 3430 (1966).]

6.6. The oxidation of iodide by hydrogen peroxide

$$H_2O_2 + 2H^+ + 2I^- \longrightarrow 2H_2O + I_2$$

proceeds by two pathways leading to the combined kinetics

$$\frac{-d(H_2O_2)}{dt} = \frac{d(I_2)}{dt} = k_1(H_2O_2)(I^-) + k_2(H_2O_2)(I^-)(H_+)$$

The ionic strength affects the rate constants at 25 °C as follows:

I	k_1, M^{-1} sec^{-1}	k_2, M^{-2} sec^{-1}
0.000	0.658	19.0
0.0207	0.663	15.0
0.0525	0.670	12.2
0.0925	0.679	11.3
0.1575	0.694	9.7
0.2025	0.705	9.2

Test Eq. (6.58) with the data. Is the agreement satisfactory in comparison with the reactions in Fig. 6.2? [F. Bell, R. Gill, D. Holden and W. F. K. Wynne-Jones, *J. Phys. Chem.* **55**, 874 (1951).]

6.7. Reaction (6.65) was studied by photylizing I_2^{131} in the presence of vinyl iodide. The mechanism is

$$hv + I_2^* \longrightarrow 2I^*$$
$$I^* + C_2H_2I_2 \longrightarrow C_2H_2II^* + I$$
$$2I^*(\text{or } I) \longrightarrow I_2^*$$

where I^* denotes I^{131}. Use the steady-state approximation to derive the differential rate expression. How could the absolute rate constants be obtained (see Appendix A.3)? (From Reference 6.)

7

Empirical Correlations

and Linear

Free Energy

Relationships

THE FOREGOING CHAPTERS dealt chiefly with the theoretical bases of reaction kinetics, but the vast majority of reactions are too complicated to be described by the theoretical apparatus so far devised, unless one is willing to settle for agreement to no better than several orders of magnitude. This problem has been circumvented by the development of a number of correlations between kinetic parameters (rate constants, activation energies, etc.) and thermodynamic or structural characteristics of the reactants. Most such correlations are empirical, but they are useful because they show which property of the reactants is of greatest importance in determining the reaction rate. As a corollary, they have been useful in guiding theoretical studies.

7.1. Hydrogen Abstraction Reactions

M. Polanyi (1) suggested that in exothermic reactions of the type

$$A + BC \longrightarrow AB + C \tag{7.1}$$

where A is a constant reactant and BC is a member of a structurally similar series of reactants, the activation energy of the reaction should vary linearly with the bond-dissociation energy of AB, D(A—B). The abstraction of hydrogen atoms from hydrocarbons by methyl groups is an example of reaction (7.1).

$$CH_3 + RH \longrightarrow CH_4 + R \tag{7.2}$$

A number of Polanyi–type correlations have been tried for this reaction, and the precision of the correlation is related to the breadth of the definition of the structurally similar class. Trotman–Dickenson has shown that when RH is restricted to saturated hydrocarbons, the activation energies are given by (2)

$$E_a = 0.49\{D(R\text{—}H) - 74.3\} \tag{7.3}$$

The correlation is shown in Fig. 7.1. It is based on only five hydrocarbons, but they are the hydrocarbons with the widest variation in D(R—H): methane, ethane, propane, n-butane, and 2-methyl propane. The agreement is within experimental error in E_a, which is several tenths of a kilocalorie per mole.

The failure of Eq. (7.3) for unsaturated hydrocarbons may be demonstrated by the extreme case of toluene, for which $D(C_6H_5CH_2\text{—}H)$ is 84 kcal/mole. Equation (7.3) predicts an activation energy of 4.8 kcal/mole in contrast to an observed 8.3. As is usual in such correlations, the exceptions receive more attention than the examples that follow the relation, for the deviation often provides some understanding of the reaction. One must be very careful in searching for reasons for discrepancies because correlations such as (7.3) are empirical; consequently, any explanation will be speculative. Bearing this point in mind, one finds an obvious difference between toluene and saturated hydrocarbons in the resonance stabilization of the tolyl radical, which has no counterpart in alkyl radicals:

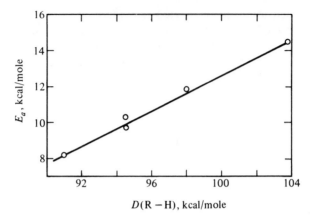

Figure 7.1. *The linear relation between the activation energy of reaction (7.2) and the bond dissociation energy* [*From Ref. (2).*]

The resonance energy gained in this fashion is responsible for the low bond-dissociation energy of toluene compared to that of ethane $[D(C_2H_5—H) = 98 \text{ kcal/mole}]$ to which it would otherwise be comparable. The lack of applicability of Eq. (7.3) to the reaction with toluene could be explained if the resonance energy of the activated complex of the reaction is very small; in other words, if the hydrogen atom must be well removed from the tolyl radical before the radical gains the resonance energy.

7.2. Free Energy Correlations

Most empirical correlations are between the logarithm of the rate constant of a reaction [proportional to the free energy of activation according to Eq. (4.26)] and a free energy change associated with one of the reactants. There are two reasons for this. First, free energies of activation are known with greater precision than activation energies. An error of 5 percent in a rate constant at room temperature corresponds to an error in $\Delta G^{\ddagger}$ of only 0.03 kcal/mole. Second, the great majority of empirical correlations concern solution reactions, and free energy correlations have been found to be more precise than either entropy or enthalpy correlations in solution.

7.3. The Brönsted Correlation

Brönsted and Pederson, in 1925, studied the base–catalyzed decomposition of nitramide

$$H_2N_2O_2 + B \longrightarrow H_2O + N_2O + B$$

and found that the reaction followed the rate law

$$-\frac{d(H_2N_2O_2)}{dt} = \{k + k_B(B)\}(H_2N_2O_2)$$

The term $k(H_2N_2O_2)$ represents the decomposition of nitramide in pure water, and $k_B(B)(H_2N_2O_2)$ corresponds to a base-catalyzed decomposition with the following mechanism:

$$H_2N_2O_2 \rightleftharpoons H-N=N{\overset{OH}{\underset{O}{}}}$$

$$B + HN=N{\overset{OH}{\underset{O}{}}} \longrightarrow BH^+ + {}^-N=N{\overset{OH}{\underset{O}{}}} \quad \text{(slow)}$$

$${}^-N=N{\overset{OH}{\underset{O}{}}} \longrightarrow N_2O + OH^- \quad \text{(fast)}$$

$$OH^- + BH^+ \longrightarrow B + H_2O$$

The logarithm of k_B varies linearly with $\log K_{HB}$, the acid-dissociation constant of the conjugate acid $HB^+ \rightleftharpoons H^+ + B$, as is shown in Fig. 7.2. Brönsted noted that his correlation generally applied to many reactions of acids

$$\ln \left(\frac{k_{HB}}{p}\right) = A + \alpha \ln \left(\frac{K_{HB}q}{p}\right) \tag{7.4}$$

as well as bases (3)

$$\ln \left(\frac{k_B}{q}\right) = A + \beta \ln \left(\frac{K_B p}{q}\right) \tag{7.5}$$

The constants p and q are statistical factors that can be found by inspection of the structure of the acid or base. The factor p is the number of equivalent protons on the acid and q is the number of equivalent sites for accepting a proton on the base. Thus for acetic acid and acetate ion, p is 1 and q is 2; for dicarboxylic acids, p is 2 and q is 2. The use of p and q is an attempt to reduce the rate constant and dissociation con-

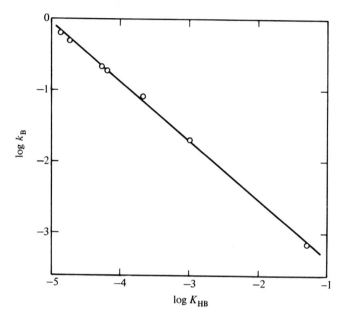

Figure 7.2. *The effect of acid strength of the conjugate acid on the base catalyzed decomposition of nitramide.*

stant to a unit acid.* If there are two acid sites in a molecule, it will affect both the rate constant and the acid-dissociation constant of an acid-catalyzed reaction, whereas if there are two potential acid sites in the conjugate base, this will affect only K_{HB}. The opposite reasoning applies to base-catalyzed reactions.

In Eqs. (7.4) and (7.5), α and β are constants varying between 0 and 1, depending on the reaction. The equations are equivalent to writing

$$\Delta G^{o\ddagger} = \alpha \, \Delta G^{o} + \text{constant} \tag{7.6}$$

that is, the free energy of activation varies linearly with the free energy of ionization of the acid.

Many attempts to derive Eqs. (7.4) and (7.5) from more basic postulates have been made with little success. It seems reasonable that if the structure of the acid is comparable in a series of acids—for example, if the series consists of aliphatic carboxylic acids—then the free energy of ionization of the acid might be the main remaining variable affecting the reaction rate. In such a case, it is apparent that for small changes in ΔG^{o}, $\Delta G^{o\ddagger}$ will be proportional to ΔG^{o}. It is not apparent what constitutes a "small" change in ΔG^{o}.

*See page 101 for a general form of the statistical factor.

the equation of substituted pentaaminecobalt complexes

$$Co(NH_3)_5X^{2+} + H_2O \longrightarrow Co(NH_3)_5(H_2O)^{3+} + X^- \tag{7.11}$$

the rate constant is proportional to the equilibrium constant for the reaction (7):

$$\log_{10}k = -5.67 + 1.0\log K_{eq} \tag{7.12}$$

Formally, this type of correlation is identical with the Brönsted correlation. Reaction (7.11) could be separated into two steps (not neccessarily mechanistic) for purposes of

$$Co(NH_3)_5X^{2+} \rightleftharpoons Co(NH_3)_5^{3+} + X^- \tag{7.13}$$

$$Co(NH_3)_5^{3+} + H_2O \rightleftharpoons Co(NH_3)_5H_2O^{3+} \tag{7.14}$$

discussion. The equilibrium constant of Eq. (7.12) may be expressed as

$$K_{eq} = K_{7.13}K_{7.14}$$

and since $K_{7.14}$ is independent of the substituent X^-, Eq. (7.11) could be written

$$\log k = 1.0 \log K_{7.13} + \text{constant} \tag{7.15}$$

Reaction (7.13) would be the equivalent of the ionization of the acid in the Brönsted formulation, and the coefficient 1.0 in Eq. (7.15) is equivalent to the Brönsted α.

The magnitude of the Bronsted constants may be interpreted to furnish information about the structure of the transition state. If an acid were completely dissociated before the reaction occurred, α would be unity. It is generally assumed that values of α near unity imply that the leaving group (the conjugate base) is very loosely bound in the transition state. For instance, the X^- group in reaction (7.11) must be well removed from the reaction site in the transition state.

7.4. The Hammett Correlation

The rates of hydrolysis of substituted ethyl benzoates show a strong dependence on the nature and position in the ring of the substituent X.

The same is true of the ionization constants of the corresponding acids, the rates of esterification of the acids, and hundreds of other reactions. Hammett noted that the effects of the substituents seemed to be correlated from one reaction to another (8).

The most extensive and precise data available were on the ionization constants of substituted benzoic acids, so he defined a parameter σ as the logarithm of the ratio of the dissociation constant for a substituted benzoic acid to that of benzoic acid, measured in water at 25 °C.

$$\sigma \equiv \log \frac{K_{HB}}{K_{benzoic\ acid}} \tag{7.16}$$

The σ constant is characteristic of the particular substituent and its position in the aromatic ring. A selected list of constants is given in columns 1 (meta substituents) and 2 (para substituents) in Table 7.1.

Other reaction rate constants and equilibria were found to follow the Hammett relation

$$\log \frac{k}{k_0} = \rho\sigma \tag{7.17}$$

where k is the rate or equilibrium constant for a substituted aromatic compound, k_0 is the rate or equilibrium constant for the analogous unsubstituted compound, and ρ is a reaction constant characteristic of the reaction series being studied. An example of the relation is shown in Fig. 7.4, where the rates of hydrolysis of substituted ethyl benzoates are shown as a

function of σ for the substituent.

Hammett applied his correlation to 39 rates and equilibria. Good agreement was found in 30 cases, including such apparently diverse reactions as the rates of acid- and base-catalyzed bromination of substituted acetophenones, the reaction of substituted benzoyl chlorides with methanol, the reaction of substituted dimethylanilines with several aromatic ethers, and the reactions of substituted phenolate ions with ethylene oxide. Jaffe (9) has made an extensive survey of aromatic reactions and found that out of 371 rates and equilibria, only 26 cases deviated seriously from the Hammett relation.

The fact that a single parameter, σ, can represent the effect of a substituent in such a wide variety of reactions suggests that σ measures one property that is of major importance in all the reactions. An obvious property is the relative electron density at the reaction site. For instance, the larger the electron density near the O—H bond in benzoic acids, the less likely the acid is to ionize. As may be seen from Eq. (7.16), σ

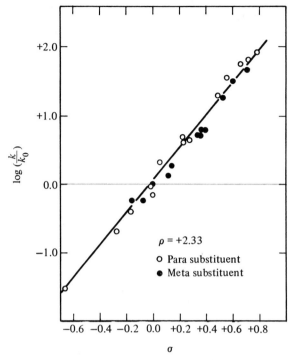

Figure 7.4. *Hammett linear free energy relationship for the rates of saponification of ethyl benzoates in 60% aqueous acetone at 25°C. [From Ref. 11.]*

decreases with decreasing K_{HB}; thus there is an inverse relation between σ and the electron density—the larger σ is, the lower the electron density at the reaction site. The largest positive values of σ in Table 7.1 are for $N(CH_3)_3^+$ substitutents, and compounds containing this group would certainly exhibit an electron deficiency at the reaction site. The constant ρ measures the sensitivity of the reaction to the electron density. A positive value for ρ indicates a reaction that is hindered by a large electron density, or a nucleophilic reaction. A negative value of ρ indicates a reaction that is accelerated by an increased electron density, or an electrophilic reaction.

Support for the identification of σ with the electron density is found in other areas of chemistry, such as a correlation of σ with the position of infrared bands. Also, Kosower (10) noted a correlation of σ with the charge transfer bands in the near UV absorption spectrum of substituted 1-alkyl pyridinium iodides. The position of the

$$RP_y^+I^- \xrightarrow{h\nu} RP_yI \tag{7.18}$$

Table 7.1. *Summary of Sigma Values*[a]

Substituent	σ_p	σ_m	σ_p^+	σ_m^+	σ_I	σ_R^p	σ_R^m
NH_2	−.66	−.15	–	–	+.10	−.76	−.25
OH	−.35	+.08	–	–	+.25	−.60	−.17
OCH_3	−.26	+.08	−.76	+.05	+.25	−.51	−.17
F	+.08	+.35	−.07	+.35	+.52	−.44	−.17
SCH_3	−.05	+.15	−.60	+.16	+.25	−.30	−.10
Cl	+.23	+.37	+.11	+.40	+.47	−.24	−.10
Br	+.23	+.39	+.15	+.41	+.45	−.22	−.06
$(CH_3)_3C$	−.16	−.11	−.26	−.07	−.07	−.09	−.04
I	+.28	+.35	+.14	+.36	+.39	−.11	−.04
CH_3	−.16	−.07	−.31	−.06	−.05	−.11	−.02
C_6H_5	−.01	+.06	−.18	+.11	+.10	−.11	−.04
H	0.00	0.00	0.00	0.00	0.00	0.00	0.00
$N(CH_3)_3^+$	+.86	+1.00	+.41	+.36	+.86	0.00	+.14
$SOCH_3$	+.56	+.53			+.52	+.04	+.01
CF_3	+.50	+.42	+.61	+.52	+.41	+.09	+.01
CN	+.68	+.62	+.66	+.56	+.58	+.10	+.04
$C_2H_5O_2C$	+.43	+.36	+.48	+.37	+.32	+.11	+.04
CH_3SO_2	+.71	+.65	–	–	+.59	+.12	+.06
CH_3CO	+.43	+.35	–	–	+.28	+.15	+.07
NO_2	+.79	+.71	+.79	+.67	+.63	+.16	+.08

[a]The values of σ_p, σ_m, σ_I, σ_R^p, and σ_R^m are from R. W. Taft, Jr., N. C. Deno, and P. S. Skell, *Ann. Rev. Phys. Chem.* **9**, 287 (1958). The values of σ_p^+ and σ_m^+ are from H. C. Brown and L. M. Stock, *J. Am. Chem. Soc.* **84**, 3298 (1962) and H. C. Brown and Y. Okamoto, *ibid.* **80**, 4979 (1958).

wavelength maximum varies with the substituent in the pyridinium ring according to

$$\frac{E_T}{2.303RT} = -13.4\sigma$$

where E_T is the energy of the wavelength shift relative to the unsubstituted compound.

The σ constant, as it is defined by Eq. (7.16), cannot depend on solvent composition or temperature, but this restriction is arbitrary. The question is whether or not other values of σ in different solvents and at different temperatures could be found which would result in better correlations.

Hammett demonstrated that Eq. (7.16) seemed sufficient to correlate rates in various solvents and at all temperatures. Jaffe's more extensive study showed that the assumption of temperature independence was satisfactory inasmuch as the magnitude of the errors in the correlation were independent of temperature. He did find a significant improvement in correlation if a few σ values were allowed to vary with solvent. The most extreme cases were m–OH, for which σ is $+0.12$ in water and -0.13 in ethanol, and p-OH, for which σ is -0.33 in water and -0.44 in ethanol. The hydroxy group would be the most susceptible to hydrogen bonding effects.

Note that the display of the data for the catalysis of the dehydration of acetaldehyde hydrate by substituted benzoic acids (filled circles, Fig. 7.3) is essentially a Hammett plot as well as a Brönsted plot. The abscissa can be converted to σ by subtracting log (q/p), 0.30 for benzoic acids, and adding the pK_{HB} of benzoic acid, 4.19. The Hammett σ is identical to the Brönsted α for the reaction [0.54 from Eq. (7.8)]. This similarity illustrates the close relationship between the two correlations.

7.5. *Other Sigma Correlations*

A number of attempts have been made to improve the correlation between substituent constants and reactivities, but this step can be done only by sacrificing generality or introducing additional constants. Brown and co-workers (11) have shown that electrophilic reactions can be better described by the relation

$$\log \frac{k}{k_0} = \rho\sigma^+ \tag{7.19}$$

where σ^+ is defined by the rate of solvolysis of substituted 2-chloro-2-phenyl propanes in 90 percent acetone, 10 percent water:

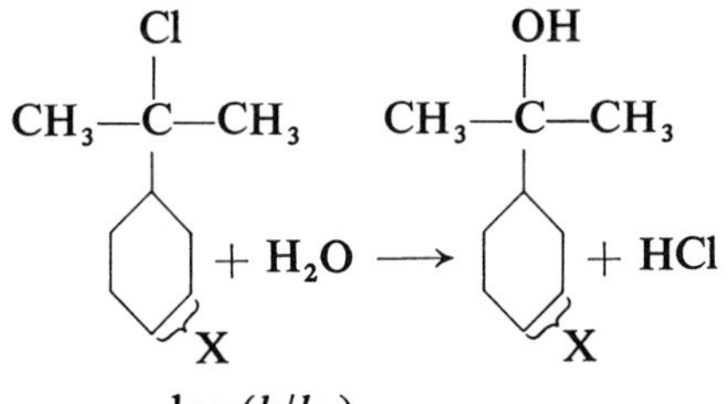

$$\sigma^+ = \frac{-\log (k/k_0)}{4.54}$$

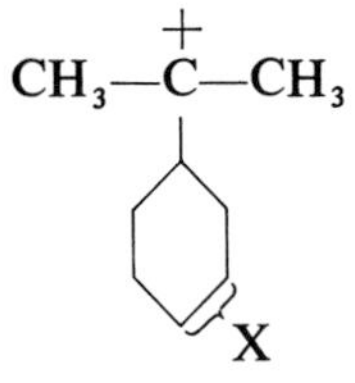

The constant -4.54 was determined as the best value of ρ based on Hammett's σ constants for meta substituents only. The hydrolysis proceeds through a carbonium ion intermediate and the leaving of the chloride ion is facilitated by a high electron density at the reaction site. This places reaction (7.20) in the category of electrophilic reactions.

An example of the correlation of an electrophilic substitution reaction, the bromination of substituted benzenes, is given in Fig. 7.5 as a function of both σ and σ^+. Some values of σ^+ are given in columns 3 and 4 of Table 7.1. Note that the values of σ^+ for para substituents are generally several tenths lower than σ.

Taft and co-workers (12) have suggested that the inductive and reso-

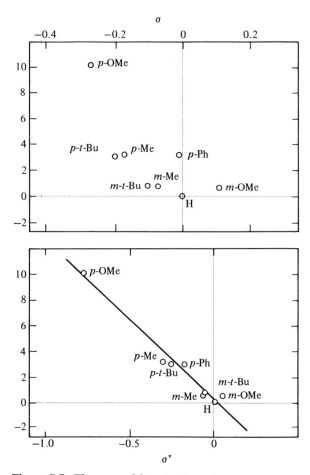

Figure 7.5. *The rate of bromination of monosubstituted benzenes in acetic acid as a function of σ(upper figure) and σ^+ (lower figure).* [*From Ref. 10.*]

nance components of the effect of substituents can be separated experimentally. The basic assumptions are that the two effects combine linearly

$$\log\left(\frac{k}{k_0}\right) = I + R$$

and that the inductive effect can be expressed by a corresponding σ_I which is characteristic of the substituent and is the same for both meta and para substitution. Roberts and Moreland measured the effect of substituents on the ionization constant K_{HB} of 4-substituted bicyclo[2.2.2]-octane-1-carboxylic acids (13):

$$X—C\overset{\displaystyle C—C}{\underset{\displaystyle C—C}{\diagup\diagdown}}C—COOH \qquad (I)$$

They defined a parameter σ' by

$$\sigma' \equiv \frac{\log (K_{HB}/K_0)}{1.464} \qquad (7.21)$$

where K_0 is the ionization constant for the unsubstituted acid measured in a 50 percent ethanol, 50 percent water solvent. The factor 1.464 is the Hammett ρ for benzoic acids in the same solvent, so that σ' and σ are as closely comparable as possible. The main difference between the two is that the effect of substituents in I is only by induction. Resonance effects cannot be present in the saturated compounds. The σ' values thus obtained produce an excellent correlation of rates of hydrolysis of ethyl esters of I, and of the reaction of I with diphenyldiazomethane.

Taft and co-workers measured the rates of acid- and base-catalyzed hydrolysis of substituted ethyl acetates

$$XCH_2COOEt + H_2O \longrightarrow XCH_2COOH + EtOH$$

and defined a parameter σ_I by

$$\sigma_I \equiv \frac{1}{6.23}\left[\log\left(\frac{k_B}{k_B^\circ}\right) - \log\left(\frac{k_{HB}}{k_{HB}^\circ}\right)\right] \qquad (7.22)$$

where the rates are measured in water at 25 °C and k_B° and k_{HB}° are the rates of reaction for ethyl propionate. The factor $1/6.23$ normalizes σ_I to σ' of Eq. (7.21).

Equation (7.22) is based on the idea that steric effects should cancel between the base- and acid-catalyzed hydrolyses due to the similarity of the two reactions. According to Hammett's values, base-catalyzed hydrolyses of aromatic compounds are quite sensitive to polar effects whereas acid-catalyzed rates are not. Consequently the difference used in Eq. (7.22) should be a good measure of inductive effects. The constants σ' and σ_I are found to be equal within experimental error, and hereafter will be referred to as σ_I. They are believed to represent the purely inductive component of σ.

The resonance components of σ for para and meta substituents are defined by

$$\sigma_R^p = \sigma^p - \sigma_I \quad \text{and} \quad \sigma_R^m = \sigma^m - \sigma_I \qquad (7.23)$$

where σ^p and σ^m are the experimental Hammett values for reactions in which resonance effects are anticipated. Even if the substituent does not have a direct resonance interaction with the reaction site or transition state, the resonance parameters will have a finite value because the resonance of the substituent with the ring will distribute a charge to the ortho and para positions (as in 7.19) that can affect the rate by induction and that is not included in σ_I. The σ_R^m constants will be smaller than σ_R^p because the resonance-induced charges in meta substituted compounds are further removed from the reaction site.

Selected values of σ_I, σ_R^p, and σ_R^m are given in Table 7.1. The substituents are arranged in order of increasing σ_R^p. If the resonance components of σ are mainly due to an *indirect* inductive effect, as is the case for the substituents in the table, then the ratio

$$\alpha = \frac{\sigma_R^p}{\sigma_R^m}$$

should be a constant. This ratio is approximately three for the substituents in the table, and its constancy lends support to the assumption that σ_I represents the inductive component of σ.

When this ratio is constant for a reaction series involving aromatic compounds, the value of σ can be combined with Eq. (7.23) to give a value of σ_I without the use of data from aliphatic reaction series:

$$\sigma_I = \frac{\alpha^p - \alpha\sigma^m}{1 - \alpha}$$

It is also possible to evaluate σ_I from the chemical shift in the F^{19} NMR spectra of substituted fluorobenzenes. The correlation of σ_I values found from aliphatic reactions (both the Taft method and the Roberts-Moreland method), from aromatic reactions, and from NMR studies is shown in Fig. 7.6. Many sigma constant series have been developed, each describing a particular type of reaction or effect. They are mainly useful as aids in unraveling the details of mechanisms and in guiding theoretctical development. Ehrenson (14) has demonstrated that the $\rho\sigma$ form of the Hammett relation can be derived for the resonance component from a molecular orbital model. Despite its oversimplifica-

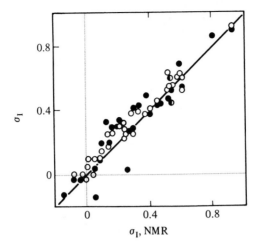

Figure 7.6. *Comparison of σ_I constants determined by the NMR chemical shifts of meta-substituted fluorobenzenes and those determined by other methods. Closed circles, aromatic reactivities; open circles, aliphatic reactivities; half filled circles, bicyclo [2.2.2] octane-1-carboxylic acids. [From Ref. 3.]*

tion, the model can be used to understand some of the largest deviation from the correlation.

7.6. The Effect of Solvent on Solvolysis Rates

The rate-determining step in many solvolysis reactions is the ionization of the substrate

$$RX \longrightarrow R^+ + X^-$$

This is followed by a rapid reaction of the carbonium ion with the solvent

$$R^+ + AB \longrightarrow RB + A^+$$

or with another anion

$$R^+ + S^- \longrightarrow RS$$

in the more general class of substitution reactions. These reactions are categorized as $S_N 1$—that is, they are substitution reactions, nucleophilic, and follow a first-order rate expression. Grunwald and Winstein found a linear free energy correlation for the rates in various solvents (15).

They define a constant Y as

$$Y \equiv \log \frac{k_{BuCl}}{k^{\circ}_{BuCl}} \tag{7.24}$$

where k_{BuCl} is the rate of solvolysis of t-butyl chloride in a given solvent and k°_{BuCl} is the rate in 80 percent ethanol. The standard temperature is 25°C, but Y is independent of temperature. The rate of solvolysis of other compounds is given by

$$\log \left(\frac{k}{k_0} \right) = mY \tag{7.25}$$

where k° is the rate in 80 percent ethanol. The solvolysis rates of neopentyl bromide in a variety of solvents are given in Fig. 7.7. The Y values are characteristic of the solvent and represent the "ionizing power" of the solvents. The value of m is characteristic of the compound and is a measure of the ease with which it can be ionized.

Kosower (10) has correlated a solvent-sensitive charge transfer band in the absorption spectrum of 1-ethyl-4-carbomethoxy-pyridinium iodide with Y values. The charge transfer spectrum is due to a process

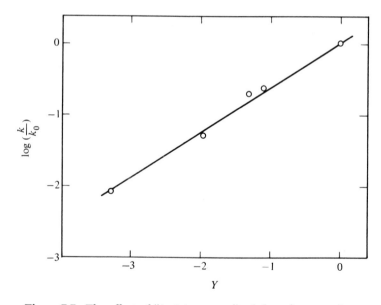

Figure 7.7. *The effect of "ionizing power" of the solvent on the rate of solvolysis of neopentyl bromide. The solvents are, from lower left to upper right, 2.5% acetic acid in acetic anhydride, ethanol, 0.1% acetic anhydride in acetic acid, methanol, 80% ethanol in water (reference solvent) [From Ref. (14)].*

such as that represented in reaction (7.18). The ground-state ion pair is very similar to the transition state in solvolysis reactions. Solvents that stabilize the ion pair will increase the energy required to reach the excited state, which is not polar. Consequently, the energy required for the transition should decrease as Y, a measure of the solvent's stabilizing effect on ion pairs, increases. The energy of the wavelength maximum of the band, Z, is given as a function of Y in Fig. 7.8.

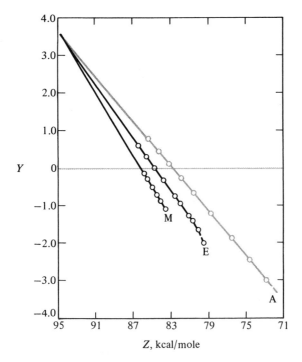

Figure 7.8. *Correlation of spectral shift (Z) with ionizing power of the solvent, Y. A, acetone-water; E, ethanol-water; M, methanol-water.* [*From E. Kosower, J. Am. Chem. Soc.* **80**, *3253 (1958).*]

7.7. Electron Transfer Reactions

Electron transfer reactions between complexed or uncomplexed metal ions are generally divided into two categories: inner sphere and outer sphere. Inner-sphere reactions involve substitution in the first coordination shell of one of the reactants and are frequently accompanied by the transfer of a ligand from one reactant to another. Outer-sphere reactions occur with the first coordination shells intact, and only an electron is moved from one reactant to another. Sutin has discussed the criteria for deciding whether a given reaction is inner sphere or outer sphere (16).

Marcus (17) has developed a theory of outer-sphere reactions that both predicts the magnitude of the rate constants and leads to a useful free energy relationship.* This correlation differs from those discussed earlier in that there is a good theoretical basis and there are no parameters to be fitted by experiment.

Transition state theory in its usual form cannot be applied to electron transfer reactions, for no bonds are made or broken. Instead, a theory based on the Franck-Condon principle is required. According to this principle, internuclear distances and nuclear momenta do not change during an electronic transition.

Since the motion of electrons is very rapid, the problem becomes that of explaining the slowness of the reactions. In fact, it is just the rapidity of electronic motion that makes the reaction slow. The large dielectric constant of water ($\mathscr{D}_s = 78$) depends on the orientation of the molecular dipoles in an electric field, a process that takes about 10^{-11} sec. If an electric field is applied that alternates in direction with a frequency greater than 10^{11} times per second, this dipole orientation lags behind the applied field. The dielectric constant drops off sharply with increasing frequency to the "optical" value, given by the square of the refractive index:

$$\mathscr{D}_0 = n^2 \cong 2$$

In the electron transfer process, the electron moves from one reactant to another in a time much shorter than 10^{-11} sec. The polarization of the solvent cannot be in equilibrium with both reactants and products at the instant of charge transfer, for the charge distribution is different in the two states. In fact, it is in equilibrium with neither, since the free energy of the system just prior to the transfer must be equal to the energy just after the transfer. The solvent rearranges to some nonequilibrium configuration before the electron transfer occurs. In agreement with the Franck-Condon principle, the nuclear configurations of the initial state (including solvent), the activated complex, and the final state are the same, but the charge distribution of the activated complex is intermediate between that of initial and final states.

The potential energy curve for an electron transfer reaction is shown schematically (in two dimensions) in Fig. 7.9. If there were *no* interaction

---

*The theory has also been applied to electrochemical reactions at electrodes, leading to a correlation of homogeneous exchange reactions and electrode reactions.

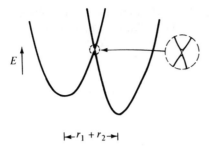

Figure 7.9. *Two dimensional analogue of the potential energy surface for an electron transfer reaction. It is postulated that there is just sufficient perturbation to split the curves for the two reactants into an upper and lower surface.*

between electronic clouds of the reactant ions, the unperturbed potential energy of each reactant would be that indicated by the broken lines. A small but finite interaction will lead to two states, represented by the upper and lower curves, and the lower curve is the adiabatic reaction path.

The problem in formulating a rate constant is to calculate the minimum free energy change $\Delta G'$ required to produce the activated complex and its associated solvent atmosphere from the initial state.

Consider the reaction

$$A^{z_A} + B^{z_B} = A^{z_A + \Delta z} + B^{z_B - \Delta z} \tag{7.26}$$

The rate constant can be expressed as

$$k_{AB} = Z_{AB} \exp\left(\frac{-\Delta G'_{AB}}{RT}\right) \tag{7.27}$$

where Z_{AB} is a collision frequency in solution (approximately 10^{11} M^{-1} sec^{-1}). Note that the definition of $\Delta G'_{AB}$ distinguishes it from $\Delta G^{\ddagger}$, since the pre-exponential factor corresponding to $\Delta G^{\ddagger}$ is kT/h rather than Z_{AB}. The free energy change $\Delta G'_{AB}$ is a sum of two components:

$$\Delta G'_{AB} = w_r + \Delta G^* \tag{7.28}$$

The first term is the work required to bring the reactants together to form a collision complex, with the solvent in dielectric equilibrium throughout the process. This is the same quantity derived in Eq. (6.30),

$$w_r = \frac{z_A z_B e^2}{\mathscr{D}_s r_{AB}} \tag{7.29}$$

except as reduced by ionic strength effects, which are significant under the usual experimental conditions. The term ΔG^* is the additional work required to transfer the electron, with the solvent in a nonequilibrium state.

We consider next a hypothetical activated complex in which a fractional charge $m \, \Delta ze$ (where Δz is the charge change on either reactant and e is the charge on the electron) is moved from reactant B to reactant A. Marcus has shown that ΔG^* is determined by this fractional charge alone, independent of the other charges on the ions. In fact, if G_0 is the free energy of formation of the pair of charges $m \, \Delta ze$ and $-m \, \Delta ze$ on the two ions in a medium of dielectric constant $\mathscr{D}_0$ and G_s is the corresponding free energy in a medium of dielectric constant $\mathscr{D}_s$, then

$$\Delta G^* = G_0 - G_s$$

The free energy G_0 is given by

$$G_0 = \frac{(m \, \Delta ze)^2}{2r_A}\left(\frac{1}{\mathscr{D}_0} - 1\right) + \frac{(-m \, \Delta ze)^2}{2r_B}\left(\frac{1}{\mathscr{D}_0} - 1\right)$$
$$+ \frac{(m \, \Delta ze)(-m \, \Delta ze)}{r_{AB}\mathscr{D}_0} \tag{7.30}$$

The first term is the work done in charging ion A of radius r_A in the medium of dielectric constant $\mathscr{D}_0$ [compare with Eq. (6.75); the -1 in the parentheses relates the free energy to that in a vacuum and is unimportant for our purposes, for it will cancel out]. The second term is the equivalent expression for ion B, while the third term is the work done in bringing the two charges from infinity to the ionic separation r_{AB} [compare with Eq. (7.29)]. For G_s,

$$G_s = \frac{(m \, \Delta ze)^2}{2r_A}\left(\frac{1}{\mathscr{D}_s} - 1\right) + \frac{(-m \, \Delta ze)^2}{2r_B}\left(\frac{1}{\mathscr{D}_s} - 1\right)$$
$$+ \frac{(m \, \Delta ze)(-m \, \Delta ze)}{r_{AB}\mathscr{D}_s} \tag{7.31}$$

and consequently

$$\Delta G^* = m^2 \lambda_0 \tag{7.32}$$

where

$$\lambda_0 = (\Delta ze)^2\left(\frac{1}{\mathscr{D}_0} - \frac{1}{\mathscr{D}_s}\right)\left(\frac{1}{2r_A} + \frac{1}{2r_B} - \frac{1}{r_{AB}}\right) \tag{7.33}$$

Combining (7.28), (7.32), and (7.33), we have

$$G'_{AB} = w_r + m^2 \lambda_0 \tag{7.34}$$

The value of m may be found by noting that the standard free energy change for the reaction is equal to the difference between $\Delta G'_{AB}$ and the corresponding quantity for the reverse reaction, $\Delta G'_{BA}$. That is,

$$\Delta G^{\circ}_{AB} = \Delta G'_{AB} - \Delta G'_{BA} \tag{7.35}$$

In the reverse reaction, the total charge to be transferred is $-\Delta ze$, and the fraction transferred in the activated complex is

$$-\Delta ze - m\,\Delta ze = -(1 + m)\,\Delta ze \tag{7.36}$$

Note that although m is a negative fraction, the choice of sign is arbitrary, for only m^2 appears in (7.34). From (7.34) we have

$$\Delta G'_{BA} = w_p + (1 + m)^2 \lambda_0$$

where w_p, the work required to bring the products together, is given by an expression like that for w_r. From (7.35) we obtain

$$\Delta G^{\circ}_{AB} = w_r + m^2 \lambda_0 - w_p - (1 + m)^2 \lambda_0$$

$$m = -\frac{\Delta G^{\circ}_{AB} + w_p - w_r}{2\lambda_0} - \tfrac{1}{2}$$

and

$$m^2 \lambda_0 = \frac{\lambda_0^2 + 2(\Delta G^{\circ}_{AB} + w_p - w_r)\lambda_0 + (\Delta G^{\circ}_{AB} + w_p - w_r)^2}{4\lambda_0} \tag{7.37}$$

Substituting this in (7.34), we obtain

$$\Delta G'_{AB} = \frac{1}{2}(w_r + w_p) + \frac{\lambda_0}{4} + \frac{\Delta G^{\circ}_{AB}}{2} + \frac{(\Delta G^{\circ}_{AB} + w_p - w_r)^2}{4\lambda_0} \tag{7.38}$$

The foregoing development did not allow the possibility of a change in the dimensions of reactants—for example, a rearrangement of molecules in the inner coordination spheres. This step would introduce an additional work term, which is essentially the potential energy of a vibrational displacement. Marcus has developed the theory for this case and finds

$$\Delta G'_{AB} = w_r + m^2 \lambda \tag{7.39}$$

$$m^2 \lambda = \frac{\lambda}{4} + \frac{(\Delta G^{\circ}_{AB} + w_p - w_r)}{2} + \frac{(\Delta G^{\circ}_{AB} + w_p - w_r)^2}{4\lambda} \tag{7.40}$$

where

$$\lambda = \lambda_0 + \lambda_i \tag{7.41}$$

and

$$\lambda_i = \Sigma_j \frac{k_j^r k_j^p (\Delta q_j)^2}{k_j^r + k_j^p} \tag{7.42}$$

In this expression, k_j^r and k_j^p are the force constants for the jth vibrational mode in the reactant and product and Δq_j is the change in the associated normal coordinate in going from reactant to product. The more complete theory is analogous to the earlier development, as may be seen by comparing Eqs. (7.39) and (7.40) with (7.34) and (7.37).

The difficulty in estimating λ_i (in no case are the required force constants and q_j's known) leads to uncertainties in estimating k_{AB} of a factor of about one thousand, thereby diminishing the usefulness of the theory for a priori calculations. The principal use of the theory arises from the fact that it relates the rate of reaction (7.26) to the rates of the corresponding isotopic exchange reactions

$$A^{z_A} + A'^{z_A + \Delta z} = A^{z_A + \Delta z} + A'^{z_A} \tag{7.43}$$

$$B^{z_B} + B'^{z_B - \Delta z} = B^{z_B - \Delta z} + B'^{z_B} \tag{7.44}$$

Apart from very small isotope effects (Section 4.11), ΔG° is zero for each exchange reaction, and also $w_r = w_p$. Therefore, for reactions (7.43) and (7.44),

$$\Delta G'_{AA} = \left(\frac{\lambda_A}{4}\right) + w_A$$

and

$$\Delta G'_{BB} = \left(\frac{\lambda_B}{4}\right) + w_B \tag{7.45}$$

A very good approximation for λ, which can be developed* from the defining equations (7.41), (7.42), and (7.33), is

$$\lambda = \tfrac{1}{2}(\lambda_A + \lambda_B) \tag{7.46}$$

where λ_A and λ_B are the values for the two exchange reactions. Equation (7.46) is in error by less than 2 percent for any reasonable values of the radii. If the work terms w_r, w_p, w_A, and w_B are neglected, Eq. (7.39) can be written

$$\Delta G'_{AB} = \frac{1}{2}\left[\frac{\lambda_A}{4} + \frac{\lambda_B}{4} + \frac{\Delta G^\circ_{AB}}{2} + \frac{(\Delta G^\circ_{AB})^2}{\lambda_A + \lambda_B}\right] \tag{7.47}$$

*If it is assumed that $r_A \simeq r_B$ and that $r_{AB} \simeq r_A + r_B$, then a good approximation to Eq. (7.33) is

$$\lambda_0 \simeq \frac{(\Delta ze)^2}{4}\left(\frac{1}{r_A} + \frac{1}{r_B}\right)\left(\frac{1}{\mathscr{D}_0} - \frac{1}{\mathscr{D}_s}\right)$$

Since the radii for ions of the same element in different valence states are very similar,

$$\lambda_{0,A} \simeq \frac{(\Delta ze)^2}{2r_A}\left(\frac{1}{\mathscr{D}_0} - \frac{1}{\mathscr{D}_s}\right)$$

and

$$\lambda_{0,B} \simeq \frac{(\Delta ze)^2}{2r_B}\left(\frac{1}{\mathscr{D}_0} - \frac{1}{\mathscr{D}_s}\right)$$

Hence

$$\lambda_0 = \tfrac{1}{2}(\lambda_{0,A} + \lambda_{0,B})$$

The corresponding equation for λ_i holds rigorously.

Then from (7.45) and (7.27),

$$\frac{\lambda_A}{4} = \Delta G'_A = -RT \ln\left(\frac{k_A}{Z}\right)$$

with similar expressions for λ_B and λ; also, by definition,

$$\Delta G^\circ_{AB} = -RT \ln K_{AB}$$

Substituting these quantities in Eq. (7.47), we obtain

$$\ln k_{AB} = \tfrac{1}{2}(\ln k_A + \ln k_B + \ln K_{AB} + \ln f)$$

where

$$\ln f = \frac{(\ln K_{AB})^2}{4} \ln\left(\frac{k_A k_B}{Z^2}\right) \tag{7.48}$$

Finally,

$$k_{AB} = (k_A k_B K_{AB} f)^{1/2} \tag{7.49}$$

Thus the rate constant of an outer-sphere electron transfer reaction can be predicted from the rates of the exchange reactions and the free energy change of the reaction. The most poorly known quantity in the theory is Z, for which the order-of-magnitude estimate of $10^{11}\ M^{-1}\ sec^{-1}$ has been used. Fortunately, it enters into Eq. (7.48) only as the logarithm, so that a tenfold change in Z produces only a 10 percent change in the logarithm of f, which, in most cases, is near unity.

Some comparisons of predicted and observed rate constants are given in Table 7.2, from which it is seen that satisfactory agreement is found

Table 7.2. *Rate Constants for Some Electron Transfer Reactions*[a]
(25°C, Units of $M^{-1}\ sec^{-1}$)

Reactants	k_A	k_B	K_{eq}	k_{calc}	k_{obs}
$Ce(IV) + Fe(CN)_6^{4-}$	4.4	3×10^2	6×10^{12}	7×10^6	1.9×10^6
$Ce(IV) + Mo(CN)_8^{4-}$	4.4	3×10^4	6×10^{10}	1.3×10^7	1.4×10^7
$Mo(CN)_8^{3-} + Fe(CN)_6^{4-}$	3×10^4	3×10^2	1×10^2	2.7×10^4	3.0×10^4
$IrCl_6^{2-} + Fe(CN)_6^{4-}$	$2.3 \times 10^{5\ b}$	3×10^2	2×10^4	8×10^5	3.8×10^5
$IrCl_6^{2-} + Mo(CN)_8^{4-}$	2.3×10^5	3×10^4	2×10^2	1.0×10^6	1.9×10^6
$Ce(IV) + Fe^{2+\ c}$	4.4	4.0	1×10^{12}	6×10^5	1.0×10^3
$Co(III) + Fe^{2+}$	5	4.0	2×10^{18}	6×10^7	42

[a]From R. J. Campion, N. Purdie, and N. Sutin, *Inorg. Chem.* **3**, 1091 (1964) and G. Dulz and N. Sutin, *ibid.* **2**, 917 (1963).

[b]P. Hurwitz and K. Kustin, *Trans. Faraday Soc.* **62**, 427 (1966).

[c]M. G. Adamson, F. S. Dainton, and P. Glenworth, *Trans. Faraday Soc.* **61**, 689 (1965).

for the first five reactions and very poor agreement for the last two. Reactions of Co(III) with ferrous phenanthroline complexes show a similar discrepancy. Sutin has suggested that there may be spin restrictions in the Co(III) reactions or that the corresponding exchange reactions may be inner sphere. It is unfortunate that few comparisons with theory are available, but not many exchange reaction rate constants are in the experimentally accessible range. Campion, Purdie, and Sutin (18) have reported four other reactions involving $W(CN)_8^{4-}$ that are in relative agreement with the Marcus theory—that is, a value for the exchange rate of $W(CN)_8^{3-}$ with $W(CN)_8^{4-}$ can be assumed which satisfies all four reactions.

A different test of Eq. (7.49) may be made by studying a homologous series of reactants. Figure 7.10 shows the rate constants for the oxidation of various ferrous phenanthrolines by Ce(IV) ion and the reduction of ferric phenanthrolines by ferrous ion as a function of the equilibrium constants of the reactions. The exchange rates of Ce(IV)—Ce(III) and

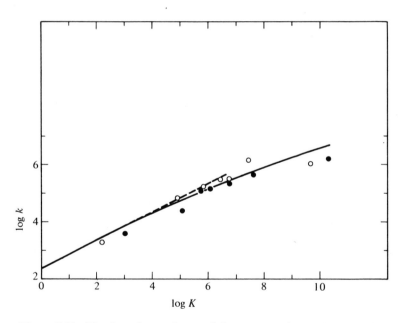

Figure 7.10. *The dependence of rate of electron transfer on the equilibrium constant of the reaction for: open circle, reduction of* Fe(phen)$_3^{3+}$ *by* Fe^{2+}; *filled circle, the oxidation of* Fe(phen)$_3^{2+}$ *by* Ce(IV). *The same substituents are studied in the two reactions, The slope and curvature are calculated according to Eq. (7.47). The dashed line is the initial slope of* $\frac{1}{2}$. *[From G. Dulz and N. Sutin,* Inorg. Chem. **2**, *917 (1963).]*

of Fe(III)—Fe(II) are 4.4 and 4.0 M^{-1} sec^{-1}; consequently, it is not surprising that the two reaction rate constant series fall on one curve. The slope of the curve and the curvature are calculated from Eq. (7.47) using a value for the Fe(phen)$_3^{3+}$—Fe(phen)$^{2+}$ exchange rate constant of 10^6 M^{-1} sec^{-1}. (Only a lower limit of 10^5 is known, but fortunately the calculation is not sensitive to the value.) A small discrepancy with the theory is apparent in the absolute values; for example, the rate constant for a reaction with $\Delta G^\circ_{AB} = 0$ would be 200, according to Fig. 7.10, but $(k_A k_B)^{1/2}$ is greater than 500 by experiment. There is agreement between the slopes of log k vs. log K curves and the Marcus theory for many reactions, including reactions of Mn(III) and Co(III) with ferrous phenanthroline complexes, and the reaction of Ce(IV) with ruthenium complexes (19).

Marcus has examined the possibility of extending the application of Eq. (7.49) to elementary reactions, such as (7.10), which do not involve electron transfer and in which bonds are broken or formed, even though such reactions violate the assumption of small interaction of the potential curves. Equation (7.49) is formally similar to the Brönsted relation. The slope of the curve in Fig. 7.10 is $\partial(\ln k_{AB})/\partial(\ln K_{AB})$ or $\partial(\Delta G'_{AB})/\partial(\Delta G^\circ_{AB})$. From Eqs. (7.39) and (7.40), ignoring work terms, we have

$$\alpha = \frac{\partial(G'_{AB})}{\partial(G^\circ_{AB})} = \frac{1}{2}\left(1 + \frac{\Delta G^\circ_{AB}}{\lambda}\right)$$

where α is the Brönsted slope. If $(\Delta G'_{AB})^\circ$ is defined as $\Delta G'_{AB}$ calculated from Eq. (7.27) for a reaction in which $\Delta G^\circ_{AB} = 0$, then

$$(\Delta G'_{AB})^\circ = \frac{\lambda}{4}$$

$$\alpha = \frac{1}{2}\left[1 + \frac{\Delta G^\circ_{AB}}{4(\Delta G'_{AB})^\circ}\right] \tag{7.50}$$

Cohen and Marcus (20) have tested Eq. (7.50) for 16 proton transfer and atom transfer reactions and find that it predicts both the correct slope and curvature, wherever curvature has been observed. Thus, when applicable, Eq. (7.50) reduces the number of empirically fit parameters to one [since $(\Delta G'_{AB})^\circ$ is determined by the A of Eq. (7.5)].

References

1. M. G. Evans and M. Polanyi, *Trans. Faraday Soc.* **34**, 11 (1938).
2. A. F. Trotman-Dickenson, *Chem. and Ind.* 1965, 379.
3. J. N. Brönsted, *Trans. Faraday Soc.* **24**, 630 (1928).
4. R. P. Bell and W. C. E. Higginson, *Proc. Roy. Soc.* (London) **A197**, 141 (1949).

5. M. Eigen, in *Proceedings of the Fifth Nobel Symposium*, S. Claesson (Ed.), Interscience Publishers, New York, 1968.
6. E. Grunwald and A. Y. Ku, *J. Am. Chem. Soc.* **90**, 29 (1968).
7. C. H. Langford, *Inorg. Chem.* **4**, 265 (1965).
8. L. P. Hammett, *J. Am. Chem. Soc.* **59**, 96 (1937).
9. H. H. Jaffe, *Chem. Revs.* **53**, 191 (1953).
10. E. Kosower, in *Progress in Physical Organic Chemistry*, S. G. Cohen, A. Streitweiser, and R. W. Taft (Eds.), Interscience Publishers, New York, 1565, vol. 3, p. 81.
11. H. C. Brown and Y. Okamoto, *J. Am. Chem. Soc.* **80**, 4979 (1958).
12. R. W. Taft, Jr., *J. Phys. Chem.* **64**, 1805 (1960). Earlier references are cited herein.
13. J. D. Roberts and W. T. Moreland, *J. Am. Chem. Soc.* **75**, 2167 (1953).
14. S. Ehrenson, in *Progress in Physical Organic Chemistry*, S. G. Cohen, A. Streitweiser, and R. W. Taft (Eds.), Interscience Publishers New York, 1964, vol. 2, p. 195.
15. E. Grunwald and S. Winstein, *J. Am. Chem. Soc.* **70**, 846 (1948).
16. N. Sutin, *Ann. Rev. Phys. Chem.* **17**, 119 (1966).
17. R. A. Marcus, *J. Phys. Chem.* **67**, 853 (1963); *J. Chem. Phys.* **38**, 1858 (1963).
18. R. J. Campion, N. Purdie, and N. Sutin, *Inorg. Chem.* **3**, 1091 (1964).
19. For a review, see N. Sutin, *Accounts Chem. Res.* **1**, 225 (1968).
20. R. A. Marcus and A. O. Cohen, *J. Phys. Chem.* **72**, 4249 (1968).

Problems

7.1. Test the values of $k_{6.63}/k_{6.64}$ for aromatic additives (Table 6.4) for a correlation with σ_m. What conclusion can you form concerning the nature of the chlorine atom complex?

7.2. Rate constants have been measured for the following reactions:

Reaction	K_{AB}	$k_{obsd}, M^{-1} \sec^{-1}$
$IrCl_6^{2-} + W(CN)_8^{4-}$	3.9×10^6	6.1×10^7
$Mo(CN)_8^{3-} + W(CN)_8^{4-}$	2.5×10^4	5.0×10^6
$Fe(CN)_6^{3-} + W(CN)_8^{4-}$	2.3×10^2	4.3×10^4

Compare these with calculated values, using exchange rate constants from Table 7.2 and assuming the exchange rate constant for $W(CN)_8^{3-} + W(CN)_8^{4-}$ to be $6 \times 10^4 \ M^{-1} \sec^{-1}$. (From Reference 18.)

7.3. Carboxylate anions are stabilized by resonance forms in which the negative charge is shared by the two equivalent oxygen atoms. In oxime anions the negative charge is confined to the oxygen atom, and in nitroparaffin anions the negative charge is completely removed from the carbon and resides on the nitro group. Devise an explanation for the relative behaviour of these three acid types observed in Fig. 7.3.

8

Complex Reactions

THE GREAT MAJORITY of reactions occurring in nature are complex—
that is, they are composed of several elementary reactions. Some of
these systems are well-understood and have been used to obtain information on elementary reactions, as has been discussed earlier in this
book. Other reactions are too complex for this purpose at present, but
kinetic studies have often been useful in elaborating the mechanisms of
these reactions. Two such reactions will be discussed in this chapter.

8.1. The α-Chymotrypsin Catalysis of Hydrolytic Reactions

Enzymes differ from other catalysts in that they are proteins produced
by living organisms. Enzyme catalysis is not an elementary reaction, but
proceeds through the formation of at least one complex. The simplest
scheme for enzyme reactions is the Michaelis-Menten mechanism
discussed in Chapter 1:

$$E + S \underset{k_{-1}}{\overset{k_1}{\rightleftharpoons}} ES \xrightarrow{k_2} E + P$$

Throughout this chapter E will represent the enzyme, S the substrate, ES and ES' complexes, and P or P_1 and P_2 products. The steady-state assumption was used in Chapter 1 to derive the rate equation

$$\mathcal{R} = -\frac{d(S)}{dt} = \frac{d(P)}{dt} = \frac{k_2}{(k_{-1} + k_2)/k_1 + (S)}(E)_0(S) \tag{1.46}$$

where $(E)_0$ is the total enzyme concentration,

$$(E)_0 = (E) + (ES)$$

Equation (1.46) is often written

$$\mathcal{R} = \frac{k_{cat}}{K_m + (S)}(E)_0(S) \tag{8.1}$$

where $K_m [= (k_{-1} + k_2)/k_1]$ is the Michaelis constant. If $k_2 \ll k_{-1}$, as is often the case, K_m is the dissociation constant of the complex.

The ultimate goal in studies of enzymatic reactions is to determine which parts of the enzyme are directly involved in the reaction, to find the role of the rest of the molecule, and to explain the reasons for the efficiency of the reaction. The main tools available to attain this goal are amino acid sequence analysis, crystallographic structure determination, and kinetic studies.

α-Chymotrypsin is a proteolytic enzyme—that is, it catalyzes the hydrolysis of esters, amides, and peptides. It is produced in the pancreas by the action of another enzyme, trypsin, on a precursor, chymotrypsinogen. It can be prepared in high purity by crystallization. The molecular weight is about 25,000 and it consists of 246 amino acid residues. The complete amino acid sequence has been worked out by Hartley (1).

A typical substrate for study might be N–acetyl–L–tryptophanamide,

The course of the reaction may be followed either spectrophotometrically, taking advantage of differences in the optical absorption spectra of products and reactants, or by amino acid analysis for the product,

N–acetyl–L–tryptophan. Usually initial rates of reaction are measured to avoid the neccessity of fitting data to the integrated form of Eq. (8.1). Accurate data can be obtained at less than a few percent reaction (2). If the initial reaction rate is defined as

$$\mathscr{R} = k_{obs}(E)_0$$

we have from Eq. (8.1)

$$k_{obs} = \frac{k_{cat}(S)}{K_m + (S)}$$

This equation could be converted into a form linear in (S) by taking the reciprocal of both sides. Another linear form is

$$k_{obs} = k_{cat} + \frac{k_{cat}(S)}{K_m + (S)} - k_{cat}$$

$$= k_{cat} - \frac{k_{cat}K_m}{K_m + (S)}$$

$$= k_{cat} - K_m \frac{k_{obs}}{(S)} \tag{8.2}$$

A plot of k_{obs} vs. $k_{obs}/(S)$ yields k_{cat} as the intercept and K_m as the slope. Some typical data obtained by Himoe et al. (2) are given in Fig. 8.1.

N-acetyl-L-tryptophanamide is a *specific* substrate for α-chymotrypsin; that is, the hydrolysis is very rapid. L–tryptophan esters are also specific substrates. A comparison of the hydrolysis rates of specific and nonspecific substrates is given in Table 8.1. Specific substrates react at rates several orders of magnitude larger than nonspecific substrates. (The slight difference in pH makes little difference in this region.) The large difference in reaction rate is not due to electrostatic effects on the reaction, since p–nitrophenyl–D–tryptophan reacts several orders of magnitude slower than the L-compound. Rather, the specificity is due to the structure of the substrate.

The acid and alcohol products do not come off the enzyme in one step. With some substrates, stable acyl-enzyme complexes may be isolated. p-Nitrophenyl acetate, for example, produces a stable acetyl chymotrypsin. Amino acid sequence analysis indicates that the acetyl group is linked to serine–195—that is, the 195th amino acid residue in the chain (5). Thus the mechanism of enzyme action is

$$E + R_1\overset{\overset{\textstyle O}{\|}}{-OCR_2} \underset{}{\overset{K_s}{\rightleftharpoons}} ES \xrightarrow{k_2} \underset{+\ R_1OH}{ES'} \xrightarrow{k_3} E + R_2\overset{\overset{\textstyle O}{\|}}{-COH} \tag{8.3}$$

where ES′ is an acyl enzyme. Applying the steady-state assumption to the concentration of ES′ and assuming equilibrium in the first step

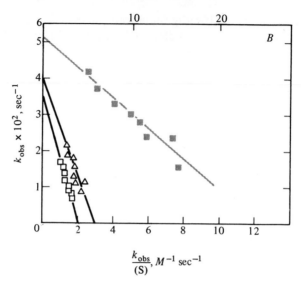

Figure 8.1. *Test of Michaelis-Menten mechanism for α-chymotrypsin-catalyzed hydrolysis of N-Acetyl-L-tryptophanamide. Filled square, pH 7.9, $(E)_0 = 5 \times 10^{-6}$ M; open square, pH 9.6, $(E)_0 = 5 \times 10^{-6}$ M; open triangle, pH 9.6, $(E)_0 = 2 \times 10^{-5}$ M. [From Ref. 2.]*

Table 8.1. *Hydrolysis Rates for Specific and Nonspecific Substrates*

Substrate	pH	k_{cat} sec^{-1}	K_m, M	Type	Reference
p-Nitrophenyl trimethylacetate	8.2	0.37	1.6×10^{-3}	Nonspecific	3
p-Nitrophenyl-L-tryptophan	7.0	31.	$2. \times 10^{-6}$	Specific	4
p-Nitrophenyl-L-phenylalanine	7.0	73.	1.1×10^{-4}	Specific	4

$$\mathcal{R} = k_3(ES') = k_2(ES) \tag{8.4}$$

$$K_s = \frac{(E)(S)}{(ES)} \tag{8.5}$$

$$(E) + (ES) + (ES') = (E)_0$$

and so, from Eqs. (8.4) and (8.5),

$$K_s\frac{(ES)}{(S)} + (ES) + \frac{k_2}{k_3}(ES) = (E)_0$$

$$(ES) = \frac{(E)_0}{K_s/(S) + k_2/k_3 + 1}$$

$$\mathscr{R} = \frac{k_3 k_2 (S)(E)_0}{k_3 K_s + k_2 (S) + k_3 (S)}$$

or

$$\mathscr{R} = \frac{k_{cat}}{K_m + (S)}(S)(E)_0 \tag{8.6}$$

where

$$k_{cat} = \frac{k_2 k_3}{k_2 + k_3} \qquad K_m = \frac{k_3 K_s}{k_2 + k_3} \tag{8.7}$$

Note that Eq. (8.6) is identical with Eq. (8.1).

Even though acyl-enzyme intermediates are not directly observed in reactions of specific substrates, their presence can be inferred from kinetic evidence. Michaelis constants for several esters are given in Table 8.2. The values of k_{cat} are seen to be essentially independent of the alcohol portion of the ester. This fact is not to be expected if the ester bond is broken in the rate-determining step. For comparison, the ester bond is broken in the rate-determining step of hydroxide ion-catalyzed hydrolysis of esters, and the rate constant for this reaction varies by a factor of 100 for the same esters. An explanation for the difference in behavior of the enzyme-catalyzed and hydroxide ion catalyzed hydrolyses is that in the former, deacylation of the enzyme is the rate-determining step. That is, if $k_2 \gg k_3$, Eq. (8.7) becomes $k_{cat} = k_3$ and $K_m = K_s k_3/k_2$. The amides of N-acetyl-L-tryptophan and N-acetyl-L-phenylalanine have values of k_{cat} that are three orders of

Table 8.2. *Reaction Constants for Specific Esters (4)*

	$k_{cat}^{sec^{-1}}$	$K_m \times 10^5\ M$
N-ACETYL-*L*-TRYPTOPHAN ESTERS		
Ethyl ester	26.9	9.7
Methyl ester	27.7	9.5
p-Nitrophenyl ester	30.5	0.2
N-ACETYL-*L*-PHENYLALANINE ESTERS		
Ethyl ester	63.1	88
Methyl ester	57.5	150
p-Nitrophenyl ester	77	2.4

magnitude smaller than the esters. Since k_3 must be the same for the amides as for the esters, the inequality must be reversed, $k_3 \gg k_2$; and so for amides $k_{cat} = k_2$, and the acylation step is rate-determining.

The nucleophilic nature of the reaction has been established by measuring the hydrolysis rates of several substituted phenyl esters. The hydrolyses of substituted phenyl trimethylacetates are characterized by a very slow deacylation step (3). This step is so slow that the rate of production of the phenol in the acylation may be followed by considering ES′ to be a permanent product. The complex ES is considered in equilibrium with E and S:

$$\frac{(ES)}{(E)(S)} = \frac{1}{K_s}$$

$$\frac{d(ES')}{dt} = k_2(ES) = \frac{k_2}{K_s}(S)(E) \tag{8.8}$$

and the enzyme concentration (E) is related to (ES′) by

$$(E)_0 = (E) + (ES) + (ES')$$

$$= (E)\left[1 + \frac{(S)}{K_s}\right] + (ES')$$

or

$$(E) = \frac{(E)_0 - (ES')}{1 + [(S)/K_s]} \tag{8.9}$$

At long times, $(ES')_\infty = (E)_0$. Therefore, inserting Eq. (8.9) into (8.8), we have

$$\frac{d(ES')}{dt} = \frac{k_2}{[K_s/(S)] + 1}[(P)_\infty - (P)]$$

This expression can be integrated if (S) is essentially constant to give

$$\ln[(P)_\infty - (P)] = \ln(P)_\infty - \frac{k_2 t}{[K_s/(S)] + 1}$$

Thus ES′ will follow a pseudo-first-order rate expression with rate constant $k_2/[K_s/(S) + 1]$. Most of the work has been limited to substrate concentrations much lower than K_s; hence the apparent first-order rate constant is approximately k_2/K_s.

The correlation of values of k_2/K_s with the corresponding σ^- constants is shown in Fig. 8.2. The σ^- constant is a substitution constant similar to Hammett's σ (Section 7.4), but it is derived from the ionization of phenols instead of benzoic acids. The use of σ^- results in a better correlation of the limited group of nucleophilic reactions of phenols and anilines. Its use is indicated here, since the reaction products are phenols.

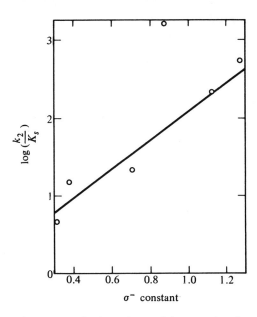

Figure 8.2. *The dependence of the rate of acylation of α-chymotrypsin on σ^- for substituted phenyl acetates. From left to right, m-acetyl, m-aldehydo, m-nitro, p-acetyl, p-aldehydo and p-nitro. [From Ref. 3.]*

The positive slope in Fig. 8.2 indicates that the reaction is nucleophilic (3). Of course, the effect could be due to either k_2 or K_s, but separate studies on three substrates capable of being studied at sufficiently high concentration to separate K_s and k_2 indicate that the main effect is on k_2. The value of Hammett's ρ for the acylation rate constant, k_2, is found to be approximately 1.4, and the value of ρ for K_s is −0.3 (3).

A separate study of the deacylation reaction indicates that it, too, is nucleophilic with a ρ of +2.1 (6).

An important clue to the composition of the active site of the enzyme is found in the pH dependence of the reactions. The variation with pH of k_{cat} and K_m for the substrate N-acetyl-L-tryptophanamide is shown in Fig. 8.3. The reaction is affected by groups with apparent pK's of 7 and 10 (for k_{cat}) and 9 (for K_m). The group with a pK of 7 affects the reactions of all substrates that have been studied (7), and consequently must pertain to some group on the enzyme. There are only three groups in the enzyme with a pK in this region: a terminal NH_3^+, which is unlikely to be involved in the catalysis, and two imidazole rings, which are incorporated into two histidine residues (8). Imidazole itself

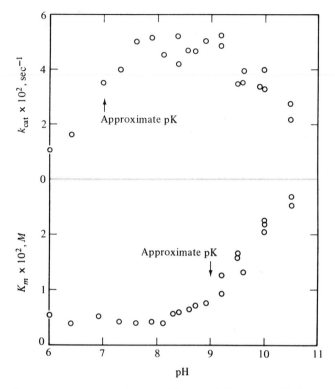

Figure 8.3. *The dependence of k_{cat} and $K_{m_{app}}$ on pH for α-chymotrypsin-catalyzed hydrolysis of N-acetyl-L-trypto-phanamide.* [*From Ref. 2.*]

$$N \quad NH + H^+ \rightleftharpoons HN^+ \quad NH$$

has a pK of 7.09 in water and 7.65 in D_2O. Bender et al. have found that the apparent pK of the group affecting k_{cat} at pH 7 also shifts by $+0.5$ pK unit in D_2O (7). Consequently, it is apparent that at least one of the histadines is implicated in the active site. Subsequently it has been found that *L*-1-tosyl-amido-2 phenylethylchloromethyl ketone reacts with α-chymotrypsin to methylate the imidazole ring attached to histadine-57 (9). The resulting methylated enzyme is inactive, which is strong evidence for the participation of this particular imidazole in the catalysis.

Thus kinetic studies and amino acid sequence analysis have combined to point to the combination of the histadine-57 residue and serine-195 residue as being part of the active site even though these residues are separated by 137 other amino acids! The crystallographic structure determination of α-chymotrypsin by Sigler et al. (10) has shown that histadine-57 and serine-195 are, indeed, adjacent to one another. A schematic representation of the structure is shown in Fig. 8.4. The structure determination also suggests an origin for the pK 9 effect. The carboxylate ion of the aspargine-194 residue is held away from the active site by formation of an ion pair with the NH_3^+ of isoleucine-16. At high pH, the NH_3^+ is neutralized and the carboxylate ion is free to rotate into the active site. Apparently this effect destroys the specificity of the enzyme, for some nonspecific substrates do not exhibit a pK effect at pH 9 (7).

A minimal description of the action of α-chymotrypsin, then, is

$$E + S \underset{}{\overset{K_s}{\rightleftharpoons}} ES \xrightarrow{k_2} ES' \xrightarrow{k_3} E + P_2 \qquad (8.10)$$

$$\Big\updownarrow \text{pK}=9 \qquad \Big\updownarrow \text{pK}=7 \qquad +$$

$$E_i \qquad ESH^+ \qquad P_1$$

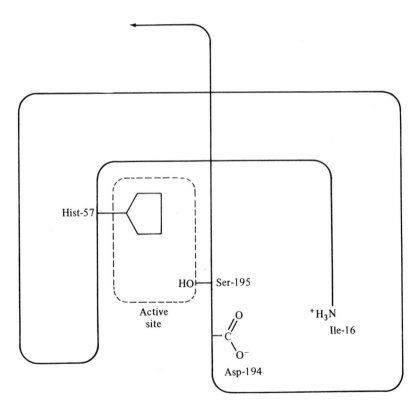

Figure 8.4. *Schematic drawing of the active site of α-chymotrypsin. [From Ref. 10.]*

where E_i is an inactive form of the enzyme, ESH^+ is an inactive complex, protonated on the imidazole ring of histadine-57, ES' is an acyl enzyme, P_1 is the alcohol or amine product, and P_2 is the acid product. The decrease in k_{cat} above pH 9 may be due to the reaction of the inactive form of the enzyme, E_i.

In the course of the studies in D_2O, it was also found that the values of k_2 and k_3 for various substrates were always a factor of 2 to 3 times larger in H_2O than in D_2O. This observation suggests that at least one bond to a hydrogen is involved in each of the two activated complexes. On the basis of the isotope effect, Bender and Kézdy (11) suggest that acylation of serine-195 takes place by simultaneous proton transfer from the hydroxyl group of serine-195 to histadine-57, and from the latter to the alcohol portion of the ester:

$$(8.11)$$

This mechanism suggests that the deacylation step, step 3 of reaction (8.10), might be considered as parallel to the inverse of the acylation step of (8.11) except that water replaces the alcohol R_1OH. Certainly the isotope effect is consistent with this assumption. The mechanism requires water to be directly involved in the activated complex of the deacylation reaction. Since water is the solvent for the reaction, the order with respect to water cannot be found without changing solvents. Bender et al. (12) have studied the deacylation of trans-cinnamoyl-α-chymotrypsin in methanol-water mixtures, and they find that the deacylation reaction is indeed first order in water and that a parallel methanolysis, first order in methanol, also occurs.

Havsteen (13) has demonstrated that the mechanism (8.10) is indeed an oversimplification. He observed another complex, which precedes the formation of ES in the mechanism:

$$E + S \rightleftharpoons ES'' \rightleftharpoons ES$$

The complex was found by using the temperature jump technique (see Appendix A, page 245) and with proflavin (3, 6-diaminoacridine) as the substrate. The rate constant for formation of ES'', 0.63×10^8 at pH 8.42 and 12 °C, is very near the diffusion-controlled limit. The conversion of ES'' to ES is probably a conformational change prior to the catalytic steps.

8.2. Chemistry of the Upper Atmosphere

The distribution of the elements of the atmosphere among the various molecular and radical species (O_2, N_2, O_3, O, etc.) is determined by photochemical processes. The problem faced by kineticists—that of describing these processes—must be solved by the study of model systems in the laboratory, from which elementary rate data are obtained, and by the application of these rate data to formulate a model atmosphere. If the composition of the model atmosphere and the actual atmosphere are in agreement, a satisfactory description is available. We shall discuss a few of the many aspects of atmospheric studies here. Some will be mentioned only qualitatively, for quantitative description is lengthy and the derived model atmospheres are as yet rarely in more than semi-quantitative agreement with the actual atmosphere.

The condition of the upper atmosphere is not constant at all times and places; there are variable winds, seasonal and daily variations of light intensity, and so on. A Standard Atmosphere has been developed for purposes of calculation, and the density and temperature of the U. S. Standard Atmosphere are given in Fig. 8.5. The atmosphere is divided into two regions, the homosphere and the thermosphere. The gross composition of the homosphere is essentially the same everywhere, 78 percent nitrogen, 21 percent oxygen, and 1 percent argon. The homosphere is further divided into the troposphere below 15 km, which will not concern us here, the stratosphere, between altitudes of about 15 km and 50 km, characterized by a rising temperature from 217 to 280 °K, and the mesosphere between 50 km and about 85 km, in which the temperature drops to the lowest point, 190 °K. The temperature of the thermosphere rises continually with altitude, and thermal equilibrium does not exist at very high altitudes.

The main chemical features of the thermosphere are the ion layers, which occur mainly above 100 km. Photoionization of N_2, O, and O_2 produces N_2^+, O^+, and O_2^+, which undergo ion-molecule reactions to form NO^+ (14).

$$N_2^+ + O \longrightarrow NO^+ + N \qquad k = 1.5 \times 10^{11} \ M^{-1} \ \text{sec}^{-1}$$

$$N_2^+ + O_2 \longrightarrow N_2 + O_2^+ \qquad k = 6 \times 10^{10}$$

$$O^+ + N_2 \longrightarrow NO^+ + N \qquad k = 1.2 \times 10^9$$

As a result, NO^+, O_2^+, and O^+ are the principal ions present. Above 100 km, the electron is the principal negative species (15), and ion recombinations are by the very rapid dissociative processes (k about $10^{14} \ M^{-1} \ \text{sec}^{-1}$)

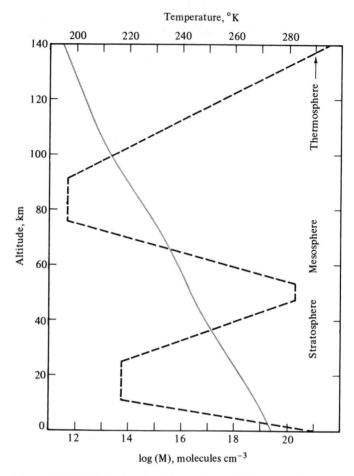

Figure 8.5. *The U. S. Standard Atmosphere. Solid line, total gas concentration; dashed line, temperature.*

$$NO^+ + e \longrightarrow N + O$$

$$O_2^+ + e \longrightarrow 2\,O$$

and the radiative process (k about $3 \times 10^9\ M^{-1}\ sec^{-1}$)

$$O^+ + e \longrightarrow O + h\nu$$

The nitrogen atoms produced in the thermosphere are replaced by oxygen atoms in the sequence

$$N + O_2 \longrightarrow NO + O \tag{8.12}$$

$$N + NO \longrightarrow N_2 + O \tag{8.13}$$

Reaction (8.12) is rather slow, with an activation energy of 7 kcal/mole, but (8.13) is extremely rapid—$k = 1.3 \times 10^{10}\ M^{-1}\ \text{sec}^{-1}$ (16). Atoms do not recombine at higher altitudes because the recombinations are termolecular processes with rate constants around $10^9\ M^{-2}\ \text{sec}^{-1}$ while the concentrations are low. For instance, the total density of molecules is about $2 \times 10^{-10}\ M$ at 150 km. Even if the gas consisted only of oxygen atoms, the rate of recombination, $\mathscr{R} = k(\text{O})^3$, would be only $10^{-20}\ M$ sec^{-1}. The lifetime of the radicals with respect to combination would be $(\text{O})/\mathscr{R}$, which is 2×10^{10} sec, or about 1000 years! Instead of recombining at high altitudes, the atoms move by diffusion and by eddies to lower altitudes (around 100 km), where the pressure is higher and recombination can occur.

Oxygen atoms provide the main reservoir of energy for the night glow of the atmosphere. One of the most conspicuous lines in the night-glow spectrum is the 5577 Å line due to the transition

$$O(^1S) \longrightarrow O(^1D) + h\nu \tag{8.14}$$

The principal source of $O(^1S)$ found in the laboratory is the three-body combination of ground-state oxygen atoms (17):

$$3\ O(^3P) \longrightarrow O_2 + O^* \tag{8.15}$$

The $O(^1S)$ (designated O^*) emit light and are also deactivated by collision,

$$\left.\begin{array}{l} O^* + O \longrightarrow \\ O^* + M \longrightarrow \end{array}\right\} \text{deactivation} \qquad \begin{array}{l} (8.16) \\ (8.17) \end{array}$$

where M is any added gas. A steady-state treatment of the concentration of O^* in reactions (8.14) to (8.17) leads to*

$$k_{15}(\text{O})^3 = k_{14}(\text{O}^*) + k_{16}(\text{O}^*)(\text{O}) + k_{17}(\text{O}^*)(\text{M})$$

But $k_{14}\ (\text{O}^*)$ is just the rate of light emission, I, so

$$\frac{k_{15}(\text{O})^3}{I} = 1 + \frac{k_{16}}{k_{14}}(\text{O}) + \frac{k_{17}}{k_{14}}(\text{M}) \tag{8.18}$$

and the various rate constants and ratios may be found by plotting $(\text{O})^3/I$ as a function of (O) or (M) [keeping either (M) or (O) constant]. Young studied the chemiluminescence in a flow system using reaction (8.13) as a source of oxygen atoms to avoid direct excitation of the oxygen atoms in the discharge (17). The N atoms were produced by a microwave discharge in N_2. The oxygen atom concentration can be monitored by the chemiluminescence produced in the pair of reactions

*The rate constant of reaction (8.15) is k_{15}, etc.

$$O + NO \longrightarrow NO_2 + h\nu \qquad\qquad (8.19)$$

$$O + NO_2 \longrightarrow NO + O_2$$

Young found Eq. (8.18) to be obeyed. Oxygen atoms and O_2 are efficient quenchers, whereas N_2 is very inefficient.

The intensity of 5577 Å emission has been measured by instrumentation on rocket flights, and from these data and the measured rate constant for reaction (8.15), the oxygen atom concentration can be computed as a function of altitude. The result is given in Fig. 8.6 along with a theoretical extrapolation of the concentration from data obtained by rocket-

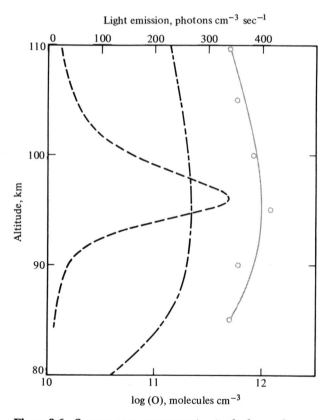

Figure 8.6. *Oxygen atom concentration in the lower thermosphere. Dashed line, light flux at 5577 Å from rocket measurements; solid line, oxygen atom concentration calculated from the light flux; dot-dash line, oxygen atom concentration from rocket-carried mass spectrometer. [From Ref. 16.]*

carried mass spectrometers above 110 km. A discrepancy between the two measurements, about a factor of five, remains to be resolved.

The largest light flux of the night glow arises from emission from vibrationally excited hydroxyl radicals. The light is in a series of bands from 5000 Å to the infrared. The maximum intensity of this emission is found at about 85 km. The hydroxyl radicals must be produced in a very exothermic reaction in order to emit this light, and the most logical candidate is the reaction (17)

$$H + O_3 \longrightarrow OH^* + O_2 \qquad \Delta H = -77 \text{ kcal/mole} \tag{8.20}$$

The rate constant for this reaction has been found to be 1.6×10^{10} $M^{-1} \text{sec}^{-1}$ at 300 °K, by measurement in a hydrogen atom flow system (16). Anlauf et al. have studied the infrared emission in a flow system and find that the OH is indeed excited (19). The intensity of the emissions indicated that the population of the vibrational states increases all the way up to the ninth level, the maximum possible with the energy available. Reaction (8.20) combines with another efficient reaction

$$OH + O \longrightarrow O_2 + H \qquad k = 3 \times 10^{10} \, M^{-1} \text{sec}^{-1} \tag{8.21}$$

to produce a chain, so that a small concentration of hydrogen atoms suffices to produce the effect. Still, in order to predict the hydroxyl radical emission in the atmosphere, the mechanism for producing H must be introduced. One possibility is

$$O(^1D) + H_2 \longrightarrow OH + H \tag{8.22}$$

where the $O(^1D)$ atoms are produced by photolysis of O_2. Unfortunately, the rate of this reaction is unknown. Hunt, in a study of a model atmosphere, uses reaction (8.22) and the photolysis of water as the source of hydrogen atoms; he predicts that the intensity of the OH emission will peak at 80 km, in agreement with the rocket observations (20).

The chemistry of the stratosphere is dominated by the presence of ozone, which reaches a maximum concentration of 10^{-8} M at 25 km, at which point it represents only seven parts per million of the total gases. There is only 5×10^{-4} g of ozone in a one-cm^2 column of the atmosphere, but this ozone is no less important to life than oxygen, for it is the only absorber of light of wavelength longer than 2425 Å (the threshold of oxygen absorption). Without ozone, the earth would be bathed in a lethal flux of ultraviolet light. The oxygen atoms, produced in the two photolyses

$$O_2 + h\nu \longrightarrow 2\,O \tag{8.23}$$

$$O_3 + h\nu \longrightarrow O + O_2 \tag{8.24}$$

react with oxygen in a termolecular reaction to produce ozone.

$$O + O_2 + M \longrightarrow O_3 + M \tag{8.25}$$

So far in this mechanism, there is no net loss of ozone, because (8.24) is followed by (8.25). The question is: What are the reactions that consume ozone? The obvious choice is

$$O + O_3 \longrightarrow 2 O_2 \tag{8.26}$$

The rate constants for reactions (8.25) and (8.26) have been measured in oxygen atom flow systems, in the thermal decomposition of ozone, and in the photolysis of ozone. The thermal decomposition was studied by Benson and Axworthy (21), who proposed the mechanism

$$O_3 + M \underset{k_{25}}{\overset{k_{27}}{\rightleftharpoons}} O_2 + O + M \tag{8.27}$$

$$O + O_3 \longrightarrow 2 O_2 \tag{8.26}$$

A steady-state treatment of the oxygen atom concentration leads to the equation

$$\frac{(M)(O_3)}{\mathscr{R}_{O_3}} = \frac{k_{25}}{2k_{26}k_{27}} \frac{(O_2)(M)}{(O_3)} + \frac{1}{2k_{27}} \tag{8.28}$$

where $\mathscr{R}_{O_3}$ is the rate of ozone disappearance. With carefully purified O_3, using the total pressure change as a measure of ozone decomposition, Benson and Axworthy found that Eq. (8.28) was obeyed. When M is either O_2 or N_2,

$$k_{27} = 2.0 \times 10^{12} \exp\left(\frac{-24{,}000}{RT}\right) M^{-1} \sec^{-1}$$

$$\frac{k_{27}k_{26}}{k_{25}} = 2.3 \times 10^{15} \exp\left(\frac{-30{,}600}{RT}\right) \sec^{-1}$$

from which

$$\frac{k_{26}}{k_{25}} = 1.1 \times 10^3 \exp\left(\frac{-6{,}600}{RT}\right) M$$

The individual rate constants can be obtained from the thermodynamics of reaction (8.27), for which the equilibrium constant, $K = k_{27}/k_{25}$, is

$$6.8 \times 10^4 \exp\left(\frac{-24{,}890}{RT}\right) M$$

Combining this with k_{27} above, we have

$$k_{25} = 3.0 \times 10^7 \exp\left(\frac{890}{RT}\right) M^{-2} \sec^{-1}$$

and from the ratio k_{26}/k_{25},

$$k_{26} = 3.4 \times 10^{10} \exp\left(-5{,}700/RT\right) M^{-1} \sec^{-1}$$

The activation energies are found as the difference between two large numbers and thus are not very precise. More direct determinations are desirable, but flow-system experiments gave very erratic results. Kaufman and Kelso found that the trouble lay in the production in the discharge tube of excited states of oxygen, which interfered with the reaction (22). When this difficulty was avoided by producing oxygen atoms by the pyrolysis of ozone at 2000 °K, they found $k_{25} = 2.7 \times 10^8 \ M^{-2} \ \text{sec}^{-1}$ at 300 °K, a value that is a factor of two larger than that determined in the pyrolysis but that is within the error limits of the equilibrium constant for reaction (8.27).

The mechanism for ozone formation in the stratosphere could be provided by reactions (8.23) to (8.26). The rates of the photolysis reactions (8.23) and (8.24) can be calculated from the solar light spectrum and optical extinction coefficients of oxygen and ozone. Steady-state assumptions are made for the oxygen atom and ozone concentrations, and it will be assumed that the time required to establish the steady-state concentrations is sufficiently short that vertical mixing can be neglected. For oxygen atoms, we have

$$2\mathscr{R}_{23} + \mathscr{R}_{24} = \mathscr{R}_{25} + \mathscr{R}_{26} \tag{8.29}$$

and for ozone

$$\mathscr{R}_{25} = \mathscr{R}_{24} + \mathscr{R}_{26}$$

The sum of these two equations gives

$$\mathscr{R}_{26} = \mathscr{R}_{23} \tag{8.30}$$

The oxygen atom concentration, from Eq. (8.29), is

$$(\text{O}) = \frac{2\mathscr{R}_{23} + \mathscr{R}_{24}}{k_{25}(\text{O}_2)(\text{M}) + k_{26}(\text{O}_3)}$$

and with Eq. (8.30), we have

$$\frac{k_{26}(\text{O}_3)(2\mathscr{R}_{23} + \mathscr{R}_{24})}{k_{25}(\text{O}_2)(\text{M}) + k_{26}(\text{O}_3)} = \mathscr{R}_{23}$$

The values of $\mathscr{R}_{24}$ and $\mathscr{R}_{23}$ calculated from the solar spectrum indicate that $\mathscr{R}_{24} \gg \mathscr{R}_{23}$, so that

$$\frac{(\text{O}_3)}{(\text{O}_2)} = \frac{k_{25}}{k_{26}} \frac{\mathscr{R}_{23}}{\mathscr{R}_{24}}(\text{M}) \tag{8.31}$$

The photolysis rates, $\mathscr{R}_{23}$ and $\mathscr{R}_{24}$, are proportional to the O_2 and O_3 concentrations,

$$\mathscr{R}_{23} = J_{23}(\text{O}_2)$$

$$\mathscr{R}_{24} = J_{24}(\text{O}_3)$$

where J_{23} and J_{24} are the reciprocals of the photolytic lifetimes. Therefore, Eq. (8.31) becomes

$$\frac{(O_3)^2}{(O_2)^2} = \frac{k_{25}J_{23}}{k_{26}J_{24}}(M)$$

(8.32)

from which the $(O_3)/(O_2)$ ratio, and hence (O_3), can be calculated. The lifetime of the ozone is longer than one day; thus J_{23} and J_{24} must be averaged over the illumination conditions of the whole day.

The results of one such calculation are given in Fig. 8.7. Although there is a qualitative similarity between the computed curve and the measured ozone concentration, the curves differ by a factor of five, which is outside experimental error. The destruction of ozone solely by reaction (8.26) is insufficient to account for the observations. It is interesting that another mode of ozone destruction appears in laboratory studies of ozone photolysis at wavelengths below 3100 Å, where $O(^1D)$

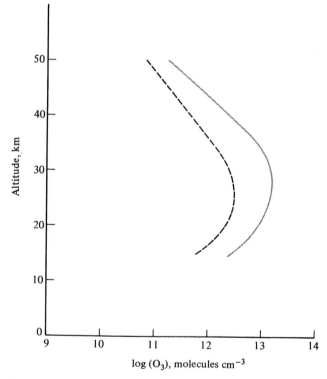

Figure 8.7. *Ozone concentration in the stratosphere. Dashed line, measurement from rocket; solid line, curve computed from Eq. (8.32). [From Ref. 19.]*

is produced. It has been suggested (23) that trace impurities (e.g., water) react with the excited oxygen atom to produce hydroxyl radicals, leading to chain decomposition of ozone:

$$O(^1D) + H_2O \longrightarrow 2\, OH \tag{8.33}$$

$$OH + O_3 \longrightarrow HO_2 + O_2 \tag{8.34}$$

$$HO_2 + O_3 \longrightarrow OH + 2\, O_2 \tag{8.35}$$

$$2\, HO_2 \longrightarrow H_2O_2 + O_2 \tag{8.36}$$

$$OH + H_2O_2 \longrightarrow HO_2 + H_2O \tag{8.37}$$

Reaction (8.35) has not been identified in any laboratory system, however, and the rate of (8.33) is not known. As a result, further study is required to decide whether this chain is involved in atmospheric ozone destruction.

The foregoing discussion has indicated some of the problems in upper atmosphere studies and the role of the kineticist in their solution. Many more reactions have been studied than were presented here, and it is apparent that new knowledge of the reactions of excited states is needed in order to develop a complete understanding of the atmosphere.

References

1. B. S. Hartley, *Nature* **201**, 1284 (1964).
2. A. Himoe, P. C. Parks, and G. P. Hess, *J. Biol. Chem.* **242**, 919 (1967).
3. M. L. Bender and K. Nakamura, *J. Am. Chem. Soc.* **84**, 2577 (1962).
4. B. Zerner, R. P. M. Bond, and M. L. Bender, *J. Am. Chem. Soc.* **86**, 3674 (1964).
5. R. A. Oosterbaan, M. van Adrichan, and J. A. Cohen, *Biochem. Biophys. Acta* **63**, 204 (1962).
6. M. Kaplow and W. P. Jencks, *Biochemistry* **1**, 883 (1962).
7. M. L. Bender, G. E. Clement, F. J. Kézdy, and H. D'A. Heck, *J. Am. Chem. Soc.* **86**, 3680 (1964).
8. H. Gutfreund, *Trans Faraday Soc.* **51**, 441 (1955).
9. E. B. Ong, E. Shaw, and G. Schoellmann, *J. Am. Chem. Soc.* **86**, 1271 (1964).
10. P. B. Sigler, D. M. Blow, B. W. Matthews, and R. Henderson, *J. Mol. Biol.* **35**, 143 (1968).
11. M. L. Bender and F. J. Kézdy, *J. Am. Chem. Soc.* **86**, 3704 (1964).
12. M. L. Bender, G. E. Clement, C. R. Gunter, and F. J. Kézdy, *J. Am. Chem. Soc.* **86**, 3697 (1964).
13. B. H. Havsteen, *J. Bio. Chem.* **242**, 769 (1967).
14. W. Fite, *Canad. J. Chem.* **47**, 1797 (1970).
15. M. A. Biondi, *ibid.*, p. 1711.

16. A. S. Vlastaras and C. A. Winkler, *Can. J. Chem.* **45**, 2837 (1967); L. F. Phillips and H. I. Schiff, *J. Chem. Phys.* **36**, 1509 (1962).
17. R. A. Young and G. Black, *J. Chem. Phys.* **44**, 3741 (1966); R. A. Young, *Canad. J. Chem.* **47**, 1927 (1970).
18. M. Nicolet, *Disc. Faraday Soc.* **37**, 7 (1964).
19. K. G. Anlauf, R. G. Macdonald, and J. C. Polanyi, *Chem. Phys. Letters* **1**, 619 (1968); J. D. McKinley, D. Garvin and M. J. Boudart, *J. Chem. Phys.* **23**, 784 (1955).
20. B. G. Hunt, *J. Geophys. Res.* **71**, 1385 (1966).
21. S. W. Benson and A. E. Axworthy, *J. Chem. Phys.* **26**, 1718 (1957).
22. F. Kaufman and J. R. Kelso, *Disc. Faraday Soc.* **37**, 26 (1964).
23. W. D. McGrath and R. G. W. Norrish, *Proc. Roy. Soc.* **A254**, 317 (1960).

General References

1. J. Westley, *Enzymic Catalysis*, Harper & Row, New York, 1969.
2. P. D. Boyer, H. Lardy, and K. Myrback (Eds.), *The Enzymes*, 2nd ed., Academic Press, New York, 1959.
3. R. D. Cadle and E. R. Allen, "Atmospheric Photochemistry," *Science* **167**, 243 (1970).
4. M. J. McEwan and L. F. Phillips, "Chemistry in the Upper Atmosphere," *Accts. Chem. Research* **3**, 9 (1970).

Appendix A

Experimental Techniques for Studying Fast Reactions

Brief discussions of the principles behind a few of the many techniques used for studying fast reactions are given in this appendix. More complete descriptions are given elsewhere [see Caldin (1), for example].

A.1. Flow Systems

Some fast bimolecular reactions may be studied by mixing the reactants, which are flowing in two separate streams, at the entrance to a tube. The reaction mixture moves down the tube with, essentially, no further mixing and the speed u with which it moves may be found by dividing the flow rate in $cm^3 \, sec^{-1}$ by the cross section of the tube. The solution passing a point a distance x along the tube left the mixer at a time x/u earlier; consequently, if the reaction is pseudo-first order and follows Eq. (1.23), the concentration of A at the distance x is given by

$$\ln (A) = \ln (A)_0 - k(B)_0 \frac{x}{u}$$

This concentration is independent of time as long as the flow is maintained, so that the analytical method used to measure (A) need not be particularly rapid. The kinetics of the reaction are studied by measuring (A) at various distances along the tube. Flow rates of 10 meters sec^{-1} are easily attainable. Consequently, reactions with lifetimes of the order of milliseconds or longer may be studied by spacing the measurements a few centimeters apart.

Experimental techniques are quite different in the liquid phase and the gas phase. Flow is established in the liquid phase by driving the plungers of two or more syringes at a known rate (see Fig. A.1). The reactants then flow into a mixing chamber, which is the most critical part of the apparatus. In the best mixing chambers, the solutions enter the flow tube tangent to the surface of the tube. An enlargement of this type of mixing chamber is shown in Fig. A.1 (the flow tube is out of the plane of the page). The use of two syringes and several entrances to the flow tube for each reactant ensures more rapid flow. The type of flow, and hence the precision of the results, depends on the flow speed. If u is too low (less than about one meter sec^{-1} for water in a 2-mm diameter tube),

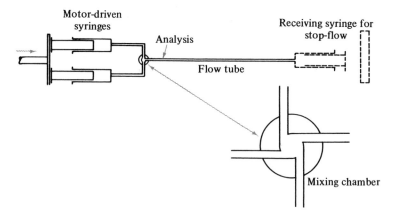

Figure A.1. *Liquid phase flow system. The dotted line portion of the apparatus is used in the stop-flow technique.*

the flow velocity at the center of the tube is greater than that near the walls (laminar flow). Turbulent flow, with a more uniform flow rate across the tube, sets in at higher flow speeds and is preferable.

The most common technique used to measure the concentration of A is absorption spectrophotometry, although conductivity, EMF measurements, ESR, and other techniques have also been used.

Most applications of flow techniques in the gas phase involve atoms or radicals (2). The reservoirs for A and B (the syringes in the liquid phase) then become bulbs filled with gas, and the flow is maintained by pumping at the exit of the flow tube. A suitable device for producing the desired unstable species (e.g., a microwave discharge) is placed between one of the reservoirs and the flow tube (usually a mixing chamber is unneccessary). The rate equation may require modification, particularly for hydrogen atoms, by addition of a term that accounts for back-diffusion of the reacting species. The appropriate expression is then

$$D_i\left(\frac{d^2 c_i}{dx^2}\right) = u\left(\frac{dc_i}{dx}\right) + \mathscr{R}$$

where D_i is the diffusion coefficient.

An important modification of the flow system, used only for liquids, is the "stopped-flow" technique, in which the flow is abruptly stopped before measurements begin. This technique is really not a flow method, but simply takes advantage of very rapid mixing; therefore the distance between the mixing chamber and the observation point is kept small. Let us assume that the concentration of some reactant is to be measured photometrically. Light of a wavelength that is partially absorbed by this compound is provided by a source, such as an incandescent lamp with an optical filter or monochromator. This light beam passes through the flow tube and the intensity is measured by a photocell or photomultiplier tube. The current from the photomultiplier is proportional to the light intensity transmitted by the solution, and thus can be related to the concentration of the reactant by the Beer-Lambert law. This current could be measured by any appropriate meter, but if the reaction is rapid, it is monitored with an oscilloscope.

Solutions of A and B are injected into the mixing chamber, usually with mechanically coupled hypodermic syringes, as in the continuous-flow method. Another syringe receives the effluent solution and is arranged so that when its plunger strikes a barrier, the flow is abruptly stopped. At the same time, a switch is struck, causing the oscilloscope to be triggered, and thus establishing the time origin. The time scale is provided by the sweep frequency of the oscilloscope; so a photograph of the oscilloscope trace is a plot of transmitted light intensity against time. These data are then used in a suitable rate expression. A distinct advantage of the stopped-flow over the continuous-flow method is the economy of reactants it permits.

A.2. Flash Photolysis and Pulse Radiolysis

These two techniques have many similar features and will be discussed together. They are most useful for studying reactions of free radicals and excited states. In flash photolysis, the sample is exposed to a very intense, short flash of light, which is absorbed principally by one component of the sample, thus providing free radicals or excited states (3). The experimental arrangement is shown in Fig. A.2. The condenser, with a capacitance of several microfarads, is charged to about 10 kV by a high-voltage power supply. A trigger signal initiates a spark in the spark gap, which then allows the rapid passage of current through the discharge lamp. With proper design, the effective resistance of the lamp will be small, thereby allowing the discharge of the condenser in a few microseconds. Very short flashes of a few nanoseconds duration can be obtained with lasers.

The concentration of the intermediates produced by the flash is followed as a function of time after the flash, generally by absorption spectrophotometry. This process can be accomplished in two ways. A smaller flash lamp, used to produce the analyzing light, is triggered to flash a short time after the large lamp. The analyzing flash enters a spectrograph, producing a photographic record of the absorption spectrum of the intermediates at that time. The second method is essentially identical with that described in the discussion of the stopped-flow method, except

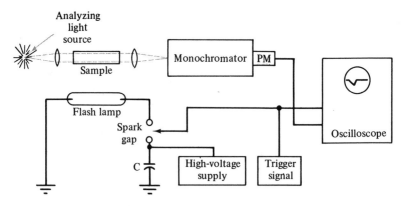

Figure A.2. *Flash photolysis apparatus.*

that the monitoring oscilloscope is triggered by the photolysis flash.

In pulse radiolysis, the flash lamp is replaced by a high-energy electron accelerator, such as a linear accelerator or a Van de Graff generator (4). Available machines produce electron beams of 0.6 to 20 MeV energy and operate at currents varying from about 0.1 to several thousand amperes for short pulses ranging upward from 10^{-9} sec. The energy of the electron beam is absorbed by all components of the sample, not just one component as was the case in flash photolysis. This disadvantage can often be overcome by the specificity of the analytical technique; for instance, a wavelength region can often be found in which only one intermediate absorbs light.

Free radical concentrations of the order of 10^{-6} M are easily obtained in flash photolysis and pulse radiolysis, and such concentrations are often observable by absorption spectrophotometry. The sensitivity of the spectrophotometric technique is limited by the intensity of the analyzing light. Approximately 10 percent of the photons striking the PM tube cause photoelectrons to be ejected from the photocathode. The statistical error in measuring the number of electrons N leaving the photocathode in time t is given by $N^{1/2}$, and this statistical fluctuation results in an inherent "noise" in the measurement. The ratio of noise to signal is $N^{1/2}/N$, or $N^{-1/2}$, when the time t is the "rise-time" of the detection system. Since N is proportional to the light flux from the analyzing lamp, the noise level varies inversely with the light intensity. With careful use of lamps of a few hundred watts power, the noise-to-signal ratio can be as low as 0.001 for rise-times of 10^{-7} sec. This ratio corresponds to a limit of detectability of the optical density measurement of 4×10^{-4}. If the extinction coefficient of the intermediate is 1000 M^{-1} cm^{-2} and the length of the light path is 4 cm, the limit of detection is 10^{-7} M.

A.3. Rotating Sector and Modulation Techniques

The rotating sector technique may be used to measure rate constants in complex reactions initiated by photolysis, if the rate of product formation does not have a linear dependence on the light intensity (5). Consider, for example, a free radical mechanism,

$$A + h\nu \longrightarrow 2R \qquad \phi I$$
$$R + S \longrightarrow R + P \qquad k_1$$
$$R + R \longrightarrow R_2 \qquad k_2$$

where ϕ is the quantum yield for production of R. The second reaction may actually occur in several steps and involve other intermediates as long as subsequent steps are rapid. Since R is regenerated in the second reaction (or reactions), a steady-state treatment of (R), under constant illumination, yields

$$\phi I = 2k_2(\text{R})^2$$

$$(\text{R}) = \left(\frac{\phi I}{2k_2}\right)^{1/2}$$

and

$$\frac{d(\text{P})}{dt} = k_1(\text{R})(\text{S}) = \left(\frac{k_1^2 \phi I}{2k_2}\right)^{1/2}(\text{S})$$

Hence the rate constant ratio, k_1^2/k_2, can be determined by measuring the quantum yield and the rate of product formation. The average lifetime τ of a radical R with respect to the recombination reaction (i.e., neglecting any interchange of radicals in the second reaction) is given by (R) divided by the rate of generation of R, ϕI:

$$\tau = \tfrac{1}{2}(k_2 \phi I)^{-1/2}$$

Now, assume that the photolyzing light passes through open sectors cut in a rotating disk such that one-half of the disk is open. If the disk rotates so fast that the light is interrupted several times within the time τ, the system will respond as if the light intensity were $I/2$. The rate becomes

$$\frac{d(\text{P})}{dt} = \frac{1}{2}\left(\frac{k_1^2 \phi I}{k_2}\right)^{1/2} = \mathscr{R}^\infty$$

If, on the other hand, the disk rotates very slowly, (R) will reach its steady-state value during illumination and approach zero during the dark period. For this case, the rate is

$$\frac{d(\text{P})}{dt} = \frac{1}{2}\left(\frac{k_1^2 \phi I}{2k_2}\right)^{1/2}(\text{S}) = \mathscr{R}^0$$

The ratio of these two rates is

$$\frac{\mathscr{R}^\infty}{\mathscr{R}^0} = \sqrt{2}$$

The reaction rate changes as the period of illumination, t_1, approaches τ, the lifetime of R. An exact integration of the rate equations, without the steady-state assumption, relates the ratio $\mathscr{R}/\mathscr{R}^0$ to the dimensionless parameter

$$\beta = \frac{t_1}{\tau} = (2k_2 \phi I)^{1/2} t_1$$

The lifetime, and hence k_2, can be determined by a comparison of the experimental results with the computed curve; k_1 can then be obtained from the ratio k_1^2/k_2 obtained with steady illumination.

The rotating sector technique may still be used even if R is not regenerated in the second reaction—that is another radical, R', is produced. The mathematical description is somewhat more complex.

An example of this technique is given in Fig. A.3 for the recombination of CF_3 radicals produced by photolysis of hexafluoroacetone. The rate measured is the production of CF_3H by hydrogen abstraction from added methane. Values of t_1 ranging from 0.01 to 1 sec were employed, and $k_{CF_3+CF_3}$ was found to be 2.3×10^{10} M^{-1} sec^{-1}.

Modulation techniques also utilize intermittent illumination of the reaction mixture with a period comparable to the lifetime of an intermediate. Direct measurement of the concentration of the intermediate is substituted for a determination of the rate of product formation. The concentration of radicals in a rotating sector experiment is very low (10^{-8} to 10^{-7} M), and the statistical noise in the detection system (page 239) obscures the signal obtained during a single light or dark period. The molecular modulation method takes advantage of the repetitive nature of the signal to increase the signal-to-noise ratio.

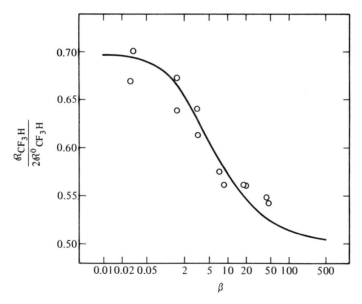

Figure A.3. *Experimental results obtained from rotating sector technique applied to the photolysis of hexafluoroacetone at 127°C. The sector ratio is two. [From P. B. Ayscough, J. Chem. Phys. 24. 944 (1956).]*

A typical plot of concentration or optical density as a function of time is given in Fig. A.4 for an intermediate following the reaction scheme

$$A + h\nu \longrightarrow 2R$$

$$R + S \longrightarrow P \qquad k_1$$

A sampling pulse, which can be obtained by suitable control of the PM circuit, is repeated with the same frequency as that of the rotating sector. A series of these pulses is then added electronically. The resulting signal is proportional to the number of sampling pulses, whereas the random noise is proportional to the square root of this number. Consequently, the relative error in the average intensity per pulse is inversely proportional to the square root of the number of pulses added together. The complete formation and decay curve of the intermediate is obtained by varying the time lag (t_1) between the onset of the photolysis period and the sampling pulse. Ultraviolet absorption spectrophotometry (6), infrared absorption spectrophotometry (7), and ESR spectroscopy (8) have been used to measure radical concentration. Radiolysis with a

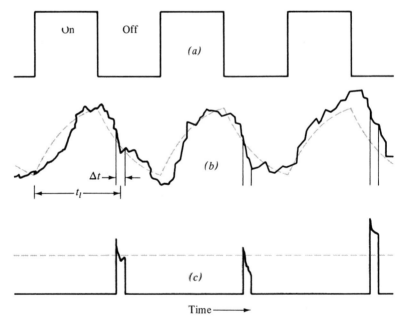

Figure A.4. *Modulation method of measuring the concentration of a short-lived reaction intermediate: (a) photolysis light intensity; (b) free radical concentration and observed signal; (c) sampling pulses obtained and summed to give the average value indicated by the dotted line.*

Van de Graaff accelerator has also been used in place of photolysis (8). An example of a modulation technique is shown in Fig. A.5.

Johnston and co-workers (7) have developed a sensitive method for determining spectra, including infrared spectra, and lifetimes of intermediates in a molecular modulation experiment. The increased sensitivity is gained not only by adding the signal from a large number of cycles as before but also by integrating the signal over a half-cycle in a way that avoids too much loss of information. Equal light and dark periods are used, and the sample flows through the reaction cell so that reactants are replenished during the dark period. The analyzing light passes through a spectrophotometer, is detected and suitably amplified. Two output signals are generated. If S_1, S_2, S_3, and S_4 are the integrals of the signals obtained for each quarter cycle (as in Fig. A.6), then

$$S_0 = S_1 + S_2 - S_3 - S_4$$
$$S_{-90°} = S_2 + S_3 - S_4 - S_1$$

As may be seen by inspection of Fig. A.6, the combination of signs for S_0 and $S_{-90°}$ characterizes the observed signal as originating from a reactant, intermediate, or product. Furthermore, the ratio of $S_{-90°}$ to S_0, in connection with the cycle length and reaction mechanism, gives

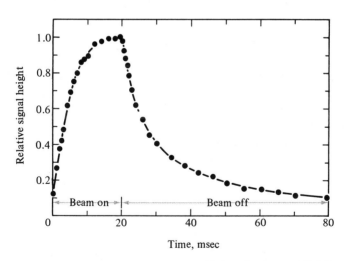

Figure A.5. *Growth and decay of ethyl radical at $-177°C$, produced by intermittent electron radiation. The relative concentration of the radical was measured by ESR. Solid curve is computed assuming the rate constant for radical recombination to be $1.8 \times 10^8 \ M^{-1} \ sec^{-1}$. The radical concentration at the peak of the curve is $1.8 \times 10^{-7} \ M$. [From Ref. 8.]*

243

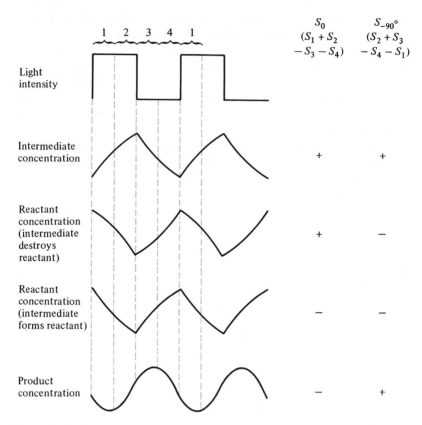

Figure A.6. *Concentration changes during molecular modulation experiment.*

the intermediate lifetime, whereas the quantity $(S_0^2 + S_{-90°}^2)^{1/2}$ is proportional to the amplitude of the variation in radical concentration.

A.4. Relaxation Techniques

Relaxation techniques have been developed extensively by Eigen and co-workers (9). These techniques involve a measurement of the time lag required to establish new equilibrium concentrations of reactants and products following the rapid application of a stress to a system in equilibrium. The position of the equilibrium can be changed by a sudden temperature jump, pressure jump, or application of an electric field. If the displacement from equilibrium is small, the approach to the new equilibrium point will be first order. Consider, for example, a recombination reaction

$$A + B \underset{k_r}{\overset{k_f}{\rightleftharpoons}} C$$

for which a, b, and c are the equilibrium concentrations under the new conditions and x is the extent to which the instantaneous concentrations (A), (B), and (C) differ from a, b, and c. The stoichiometry of the reaction gives

$$x = a - (A) = b - (B) = (C) - c$$

from which

$$\frac{-d(A)}{dt} = -\frac{d(B)}{dt} = \frac{d(C)}{dt} = \frac{dx}{dt}$$

$$\frac{dx}{dt} = k_f(A)(B) - k_r(C)$$

$$= k_f(a - x)(b - x) - k_r(c + x)$$
$$= k_f ab - k_r c - k_f(a + b) + k_r x + k_f x^2$$

By definition of equilibrium, $k_f ab - k_r c = 0$. Thus for small values of x, we have for the rate of approach to equilibrium

$$-\frac{dx}{dt} = \frac{x}{\tau}$$

The relaxation time τ is defined by

$$\tau = [k_f(a + b) + k_r]^{-1} \tag{A.1}$$

and k_f and k_r can be obtained by measuring τ as a function of $(a + b)$.

In the temperature-jump method, the temperature of the system is changed by several degrees in a time as short as 10^{-6} sec (1). The change in equilibrium constant will be given by

$$\ln \frac{K_1}{K_2} = -\frac{\Delta H^\circ}{RT^2} \Delta T$$

If $a = b \ll c$ and $\Delta H^\circ = 13$ kcal/mole (as in the dissociation of water), a temperature change of 1 °C at room temperature will shift the equilibrium concentration of A by about 3 percent.

The temperature jump may be accomplished by dissipating the energy stored in a high-voltage condenser in the solution, as in Fig. A.7. A trigger signal initiates a spark in the spark gap, thus allowing the current to flow. (This method requires that the solution contain enough electrolyte to carry the current.) The time dependence of the concentrations may be followed by the same techniques used in flash photolysis and pulse radiolysis—that is, absorption spectrophotometry or conductivity.

The shift in equilibrium may also be accomplished by a sudden pressure

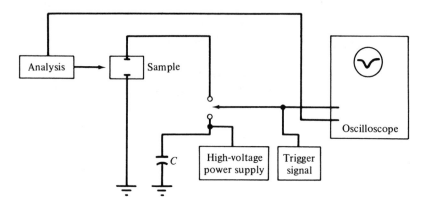

Figure A.7. *Apparatus for temperature jump technique.*

change (1). The change in the equilibrium constant is given by

$$\frac{d \ln K}{dP} = -\frac{\Delta V^{\circ}}{RT}$$

where ΔV° is the volume change in the reaction. For usual values of ΔV° (10 to 20 cc mole^{-1}), a change of 50 atm will produce a shift of a few percent in equilibrium concentrations. Still another method uses the sudden application of an electric field to shift the equilibrium of an ionic reaction through the "second Wien effect" (1).

The stress on the system can also be applied in a periodic fashion, such as pressure and temperature changes due to the passage of sound waves through the system (1). The change in the chemical composition of the system will lag behind the equilibrium composition characteristic of the instantaneous conditions and will produce an absorption of energy from the sound wave. Suppose that in the recombination reaction, y is the displacement at time t of the concentrations from a_0, b_0, c_0, the equilibrium concentrations before the passage of the sound. This displacement is

$$y = a_0 - (A) = b_0 - (B) = (C) - c_0$$

As before, the rate of change of y is

$$\frac{dy}{dt} = k_f(a_0 - y)(b_0 - y) - k_r(c_0 + y)$$

The corresponding equilibrium value of y at time t is y_{eq}; the equilibrium condition requires that

$$k_f(a_0 - y_{eq})(b_0 - y_{eq}) = k_r(c_0 + y_{eq})$$

We combine the last two equations, neglecting y^2 and y_{eq}^2, to obtain

246

$$\frac{dy}{dt} = [k_f(a_0 + b_0) + k_r](y_{eq} - y)$$

The term in brackets is the reciprocal of the relaxation time, Eq. (A.1), and therefore the last equation may be written

$$\tau\frac{dy}{dt} + y = y_{eq} \tag{A.2}$$

If the stress is due to a sound wave,

$$y_{eq} = A \sin \omega t$$

where A is the amplitude of the expected change in equilibrium concentration due to the sound wave of frequency $f = \omega/2\pi$. Then the solution of Eq. (A.2) is

$$y = \frac{A}{1 + \omega^2\tau^2} \sin \omega t - \frac{A\omega\tau}{1 + \omega^2\tau^2} \cos \omega t$$

The concentrations y_{eq} and y are shown in Fig. A.8. Note that y will increase until $y = y_{eq}$, at which point y is at a maximum ($dy/dt = 0$), and then decrease until a minimum is reached when y again equals y_{eq}. Thus y must describe the same sinusoidal form as y_{eq} except at reduced amplitude and with a phase lag given by the out-of-phase term in $\cos \omega t$. It is this out-of-phase term that leads to absorption of the sound energy. The amplitude P of a sound wave that has traveled a distance d in a fluid is given by

$$P = P_0 e^{-\alpha d}$$

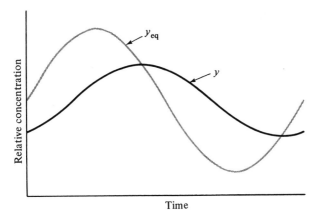

Figure A.8. *The time lag of actual concentration behind equilibrium concentration expected due to the passage of a sound wave.*

where P_0 is the amplitude at $d = 0$ and α is the absorption coefficient. The ratio α/ω is proportional to the out-of-phase coefficient, $\omega\tau/(1 + \omega^2\tau^2)$, thereby leading to

$$\frac{\alpha}{f^2} = \frac{C_1}{1 + (f/f_c)^2} + C_2$$

where $f_c = (2\pi\tau)^{-1}$, C_1 is a collection of constants, and C_2 is a constant introduced to account for background absorption that is proportional to f^2. Thus a measurement of the sound-absorption coefficient as a function of frequency leads to a value of f_c that is related to τ and hence to the reaction rate constants. A typical result obtained with this technique is given in Fig. A.9.

A variety of sound-absorption techniques covering the frequency range from about 10^3 to 10^9 Hz are available. Relaxation times ranging from 10^{-3} to 10^{-9} sec can be measured.

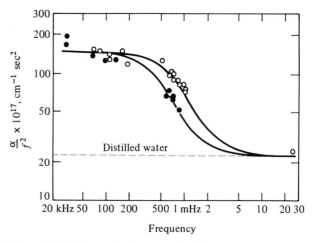

Figure A. 9. *Ultrasonic absorption* (α/f^2) *in potassium cyanide solution;* $\bullet$, *0.01 M;* $\bigcirc$, *0.05 M. The reaction causing the change in absorption is* $CN^- + H_2O \rightleftharpoons HCN + OH^-$. *The position of the absorption corresponds to* $k_f = 5.2 \times 10^4$ *sec^{-1} (pseudo-first order) and* $k_r = 3.7 \times 10^9$ *M^{-1} sec^{-1}.* [*From J. Stuehr, E. Yeager, T. Sachs, and F. Havorka, J. Chem Phys.* **38***, 587 (1963).*]

A.5. Line-Broadening Methods

Electron-spin-resonance spectroscopy (ESR) has been used as an analytical tool in flow system studies and in the molecular modulation technique, as mentioned earlier, but the line widths of ESR and NMR (nuclear-magnetic-resonance spectroscopy) spectra can be used by themselves to determine rate constants in a quite different application. The precision with which the energy difference between two spin states can be determined is limited by the uncertainty principle

$$\Delta E \tau = \frac{h}{2\pi}$$

where τ is the lifetime of the state being measured. The energy uncertainty ΔE results in a finite line width that may be expressed in frequency units, $\Delta v = \Delta E/h$. A more exact development leads to

$$\Delta v = (\pi \tau)^{-1}$$

The total line width depends on several factors, and it is only the excess over the "natural" line width that is of concern in determining rate constants. Microwave frequencies (typically about 10,000 MHz) are employed in ESR, and line broadening in the range of one megahertz to about 1000 MHz can be studied. Radio frequencies (60 MHz is common) are employed in NMR techniques, and line broadening in the range of one to 1000 Hz is observable. Both techniques are primarily used for systems at chemical equilibrium. The NMR technique is chiefly applicable to proton magnetic resonance studies and the ESR technique is applicable to free radical studies. As an example, consider the electron transfer reaction between naphthalene and naphthalenide ion (a radical ion) (10)

$$C_{10}H_8 + C_{10}H_8^{-} \longrightarrow C_{10}H_8^{-} + C_{10}H_8 \qquad (A.3)$$

The lifetime of $C_{10}H_8^{-}$ is simply the $C_{10}H_8^{-}$ concentration divided by the reaction rate, $k(C_{10}H_8)(C_{10}H_8^{-})$; thus $\tau = [k(C_{10}H_8)]^{-1}$. As the naphthalene concentration is increased, the ESR lines become broader and the line width is

$$\Delta v = \Delta v_0 + \frac{k}{\pi}(C_{10}H_8)$$

where Δv_0 is the natural line width. By this technique, the rate of reaction (A.3) for potassium naphthalenide dissolved in tetrahydrofuran was found to be $5.7 \times 10^7 \ M^{-1} \ sec^{-1}$.

References

1. E. F. Caldin, *Fast Reactions in Solution*, John Wiley & Sons, New York, 1964.
2. A. A. Westenberg, *Science* **164**, 381 (1969).
3. G. Porter, in *Techniques of Organic Chemistry*, 2nd ed., Weissberger, ed., Interscience Publishers, New York, 1963, vol. 8, part 2, p. 1055.
4. L. M. Dorfman and M. S. Matheson, in *Progress in Reaction Kinetics*, G. Porter (Ed.), Pergamon Press, New York, 1965, vol. 3, p. 239.
5. J. G. Calvert and J. N. Pitts, Jr., *Photochemistry*, John Wiley & Sons, New York, 1966, pp. 651–659.
6. E. J. Bair, J. T. Lund, and P. C. Cross, *J. Chem. Phys.* **24**, 961 (1956).
7. H. S. Johnston, G. E. McGraw, T. J. Pankert, L. W. Richards and J. van den Bogarde, *Proc. Nat. Acad. Sci. U.S.* **57**, 1146 (1967).
8. R. W. Fessenden, *J. Phys. Chem.* **68**, 1508 (1964).
9. M. Eigen, *Disc. Faraday Soc.* **17**, 194 (1954).
10. S. Weissman and R. L. Ward, *J. Am. Chem. Soc.* **79**, 2086 (1957).

Appendix B

Molecular

Beam Experiments

The actual techniques used in molecular beam studies are difficult and challenging to the experimentalist. They require a high degree of skill in vacuum engineering, mechanical design, and electronics technology. Fortunately, the principles are considerably less complex and can be easily described.

B.1. Total Scattering Cross Sections

The simplest type of experiment is the measurement of total scattering, and the principle is very much like that of measuring optical density in a spectrophotometer. There are three main chambers in the apparatus (Fig. B.1), each maintained at a very low pressure by means of suitable vacuum pumps. The requirement of low pressure ensures that particles of the incident beam will be deflected only in the scattering region, and

Source	State	Scattering	Detector
compartment	selector	compartment	compartment

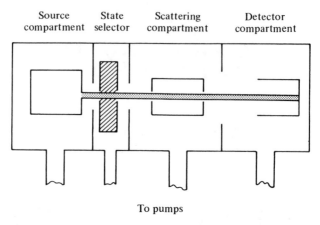

To pumps

Figure B.1. *Schematic diagram of molecular beam apparatus for measuring total scattering cross sections.*

not elsewhere in the apparatus. The first chamber contains the source of the atoms or molecules that form the primary beam. Much work has been done with alkali metal atoms, in which case the source is an oven with a narrow slit through which atoms effuse under conditions of molecular flow. The collimating slits along the beam eliminate particles moving in directions transverse to the beam axis. Particle velocities can be determined from the temperature of the oven and the Maxwell-Boltzmann equation (Appendix C). In addition, there may be a device for narrowing the inherent spread of velocities, typically a series of rotating disks that transmit only particles within a narrow velocity range. Sometimes appropriate electric fields are introduced at this point to select molecules with a particular orientation, or in specific rotational states.

The beam then enters a scattering chamber filled with the target gas at a known pressure. In more refined experiments, particularly when the angular distribution of scattering is measured, the target gas is replaced by a second molecular beam perpendicular to the primary beam. Finally, those particles that escape scattering and continue in the forward direction pass into the detection chamber, where the particle flux (particles cm^{-2} sec^{-1}) is measured, usually by a surface ionization gage or a mass spectrometer.

The experimental measurement consists of determining the flux of primary-beam particles that reach the detector when the scattering chamber is empty and when it contains scattering gas at a density of n_2 particles/cm^3. Consider a thin layer in the scattering chamber, perpendicular to the beam and with thickness dl in the beam direction, and area A in the direction normal to the beam. Within this layer, the number of scattering particles is $n_2 A\,dl$. The flux of beam particles incident on this thin layer is I_1 and that leaving is $I_1 - dI_1$. Hence the fraction scattered (i.e., not transmitted) is just dI_1/I_1, and this, in turn, must equal that fraction of the area of the thin layer that is opaque with respect to the incoming particles.

We call the effective target area of each scattering particle the *scattering cross section* (σ_{12}), so that the fraction of the total area that is opaque is simply

$$\frac{dI_1}{I_1} = \frac{-n_2\sigma_{12}A\,dl}{A} = -n_2\sigma_{12}\,dl \tag{B.1}$$

(The negative sign appears because dI_1/I_1 is the fractional *decrease* in flux.) This expression may be integrated between the limits $l = 0$ to $l = L$ (length of scattering region), to yield

$$\frac{I_1}{I_1^0} = \exp\left(-n_2\sigma_{12}L\right) \tag{B.2}$$

which has the same form as the Lambert-Beer law for light absorption.

The scattering cross section can be related to a conventional second-order rate constant in the following way. Within the volume element $A\,dl$, the total number of particles scattered per second is $A\,dI_1$, with dI_1 specified by Eq. (B.1). The flux of incident particles is related to the density n_1 and the velocity v by the relationship

$$I_1 = n_1 v \tag{B.3}$$

so that

$$AdI_1 = n_1 n_2 v\sigma_{12}(Adl) \tag{B.4}$$

We equate this result with the number scattered as expressed by a second-order rate expression

$$k_{12}C_1C_2(Adl) \tag{B.5}$$

After some juggling of units, the result is that

$$k_{12} = 10^{-3}N_{\text{Avogadro}}\sigma_{12}v$$

or

$$k_{12} = 6.02 \times 10^4\sigma_{12}v \tag{B.6}$$

When k_{12} is in liters mole^{-1} sec^{-1}, σ_{12} is in square angstroms, and v is in cm sec^{-1}.

B.2. *Differential Cross Sections*

The measurement of differential cross sections, or the angular distribution of scattered particles, usually requires that the scattering chamber be replaced by a second molecular beam. The detector is pivoted about the axis where the two beams intersect, and the flux of scattered particles at some angle ψ with respect to the initial beam is measured and compared to the flux when no scattering beam is present. In most experimental arrangements, the detector is confined to the plane of the two initial beams (i.e., the plane of the paper in Fig. B.2).

In Chapter 3 we indicated that in the center-of-mass coordinate system (CM), the distribution of scattered particles has cylindrical symmetry about the vector representing the relative velocity of the approaching particles. This situation is shown in Fig. B.3, which is a vector diagram for the case of elastic scattering when the two beams are perpendicular, as in the experimental arrangement of Fig. B.2. The vectors v_1 and v_2 represent the velocities of particles in beams 1 and 2, as measured in the laboratory coordinate system (LAB), while g is their relative velocity. The vector v_c, representing motion of the center of mass, remains unchanged in magnitude and direction. Moreover, since we are considering elastic scattering here, the vector $g = \dot{c}_1 - \dot{c}_2$ and its components are unchanged in magnitude. However, there is no restriction on the direction of the vector after scattering, g'. As indicated in Section 3.1, the angular momentum of an individual pair of interacting particles is conserved during the collision. The angular momentum vector initially lies in a plane perpendicular to g and the direction of b, the impact

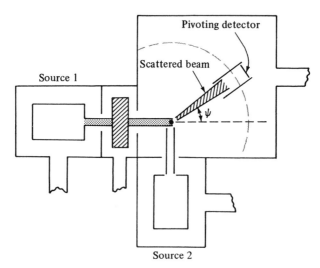

Figure B.2. *Schematic diagram of molecular beam apparatus for measuring the angular variation of scattering cross sections.*

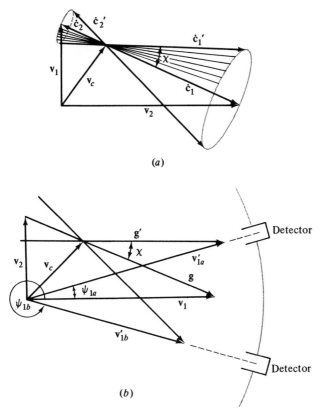

Figure B.3. (a) *Vector diagram for velocities involved in elastic scattering. In the laboratory coordinate system (LAB), the initial velocity vectors are* v_1 *and* v_2, *while* $\dot{c}_1$ *abd* $\dot{c}_2$ *are the corresponding velocity vectors in the center-of-mass coordinate system (CM). The primes indicate velocities after scattering.* (b) *The same diagram as (a) but with scattering confined to the plane of the drawing. A single scattering angle in CM coordinates* (χ) *corresponds to two angles,* ψ_{1a} *and* ψ_{1b}, *and two velocities,* v'_{1a} *and* v'_{1b}, *in LAB coordinates.*

parameter. Although the direction of **g** is fixed by the experimental conditions, the orientation of b is completely random. Therefore all orientations of **g**′ are permitted, each one corresponding to a particular direction of the angular momentum vector **L**. For a specific value of the deflection angle χ in the CM system, the vector **g**′ sweeps out a cone around the axis **g**. If the detector is confined to a plane, then, as shown in Fig. B.3(b), there are two positions of the detector (at angles ψ_{1a} and ψ_{1b}) that correspond to a single value of χ. Moreover, the velocity of the scattered particles in the LAB system has two values, v'_{1a} and v'_{1b}.

Of course, the actual measurements of the angular distribution of scattered particles are made in the LAB system, whereas kinematic theory is best expressed in the CM system. These two systems can be related (1), although doing so is not always simple. For the experimental arrangement described above, one can show that the scattering angles in the two coordinate systems are related by the expression

$$\chi = \cos^{-1}\left[\frac{(2g^2 - v_1^2 - v_1'^2 + 2v_1 v_1' \cos \psi_1)}{2g^2}\right] \tag{B.7}$$

This coordinate transformation must also be taken into account in order to obtain properly the differential cross section as a function of χ.

B.3. Inelastic Scattering

The important distinction of inelastic scattering (whether reactive or not) is that internal energy can be transformed to kinetic energy, and vice versa. The requirement of energy conservation leads to the relation

$$m_3 v_3^2 + m_4 v_4^2 - m_1 v_1^2 - m_2 v_2^2 = 2\,\Delta E \tag{B.8}$$

In this expression, 1 and 2 refer to molecules before collision and 3 and 4 refer to molecules after collision. The total mass remains constant so $m_3 + m_4$ is equal to $m_1 + m_2$. The quantity ΔE represents the total exothermicity of the collision process; in terms of bond-dissociation energies (D), and internal energies (I) of the molecules involved, it is

$$\Delta E = D_3 + D_4 - (D_1 + D_2) + I_1 + I_2 - (I_3 + I_4)$$

Again, the kinetic energy of the moving center of mass does not change, so that Eq. (B.8) may also be written in the form

$$m_3 \dot{c}_3^2 + m_4 \dot{c}_4^2 - m_1 \dot{c}_1^2 - m_2 \dot{c}_2^2 = \mu' g'^2 - \mu g^2 = 2\,\Delta E \tag{B.9}$$

From this equation we obtain

$$g' = g\left[\frac{\mu}{\mu'} + \frac{\Delta E}{E_0}\right]^{1/2}$$
$$= g\left[\frac{m_1 m_2}{m_3 m_4}\left(1 + \frac{\Delta E}{E_0}\right)\right]^{1/2} \tag{B.10}$$

Therefore the appropriate velocity vector diagram is similar to that of Fig. B.3, but with the preceeding change in the magnitude of the relative velocity vector. This vector is partitioned into the individual molecular velocities $\dot{c}_3$ and $\dot{c}_4$ according to

$$\dot{c}_3 = \frac{m_4 g'}{(m_3 + m_4)} \quad \text{and} \quad \dot{c}_4 = \frac{m_3 g'}{(m_3 + m_4)} \tag{B.11}$$

The tip of the velocity vector $\dot{c}_3$ describes a sphere centered at the CM origin and with a radius determined by Eqs. (B.10) and (B.11). Thus in a planar configuration of the two molecular beams and the detector, this vector describes a circle that is determined in size by the value of ΔE (Fig. B.4). Usually exothermic reactions are studied so that $\Delta E > 0$ and g' is greater than g. If $E \cong 0$, then this circle is small, and the LAB scattering angle ψ is well defined and near the angle between $\mathbf{v}_1$ and $\mathbf{v}_c$. As the circle becomes larger, this angle has an increasing range of allowable values, until it extends from 0 to 2π when $\dot{c}_3 = v_c$. However, a simultaneous measurement of ψ and $\mathbf{v}_3$ permits the reconstruction of g', and with the other quantities of Eq. (B.10) determined by experimental conditions, the measurement of ΔE. Results obtained in this way for the reaction $K + Br_2 \longrightarrow KBr + Br$ are shown in Fig. B.4. An additional example is given in Section 3.10.

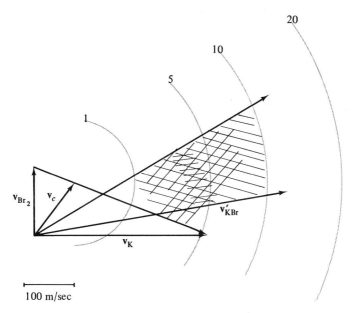

100 m/sec

Figure B.4. *In-plane reactive scattering of* KBr *from the reaction of* K *with* Br_2. *Concentric arcs represent different values of the final relative kinetic energy,* $\frac{1}{2}\mu'g'^2 = \frac{1}{2}\mu g^2 + \Delta E$ (*in kcal./mole*). *Under the experimental conditions portrayed, the internal energy of the* KBr *product is* $46\text{-}\frac{1}{2}\mu'g'^2$ *kcal/mole. The shaded area indicates the region of maximum scattering intensity, and this indicates that much of the energy of reaction appears as internal energy of the* KBr. *[Adapted from J. H. Bireley and D. R. Herschbach, J. Chem. Phys.,* **44** (*1966*).]

In many of the earlier molecular beam experiments, product velocities were not determined because of intensity problems. Even so, angular distribution measurements, together with some reasonable assumptions, provided a basis for determining the distribution of energy occurring during chemical reaction, and the type of collision leading to reaction.

Reference

1. F. A. Morse and R. B. Bernstein, *J. Chem. Phys.* **37**, 2019 (1962).

Appendix C

Summary of Some Useful Statistical Mechanical Results

Statistical mechanics provides a method for relating the average properties of an ensemble of identical systems (atoms, molecules, etc.) to the energy states available to a single member of the ensemble. The purpose of this appendix is to summarize, without detailed derivations or discussion, some results of statistical mechanics needed in Chapter 4 and subsequent chapters. Additional discussion at an elementary level may be found in physical chemistry texts (1); also, a number of textbooks devoted specifically to statistical mechanics or statistical thermodynamics are available.

C.1. Distribution Functions

Suppose that we have a large library containing N books and that we note the number of pages (x) in each book. If the library is large enough, there will be coincidences in the number of pages per book and N_1

books will have x_1 pages, N_2 will have x_2, etc. (Fig. C.1). For each value of x_i there is a quantity P_i such that

$$P_i = \frac{N_i}{N} \tag{C.1}$$

is the fraction of books with x_i pages or, alternatively, the probability that a book selected at random from the library will have x_i pages. Since the total number of books is

$$N = \Sigma_i N_i$$

evidently

$$\Sigma_i P_i = 1$$

The average number of pages per book is given in the usual way by

$$\bar{x} = \frac{\Sigma N_i x_i}{\Sigma N_i} = \Sigma \frac{N_i}{N} x_i = \Sigma P_i x_i$$

More generally, the average of any measured quantity M_i which is a function only of x is

$$\bar{M} = \frac{\Sigma N_i M_i}{\Sigma N_i} = \Sigma P_i M_i \tag{C.2}$$

If all values of N and x are sufficiently large, they can be treated as continuous rather than discrete variables. The probability P_i also becomes a continuous function, $P(x)\,dx$, defined by the expression

$$dN = NP(x)\,dx \tag{C.3}$$

where dN/N is the fraction of books containing from x to $x + dx$ pages.

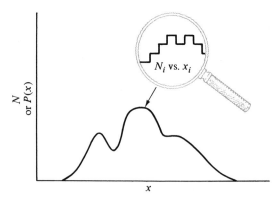

Figure C.1. *Hypothetical distribution function for a continuous variable, P(x), as a function of x. The enlarged inset shows the corresponding distribution function for discrete values of the variable, x.*

The distribution function $P(x)$ has been defined in such a way that

$$\int_{\text{all } x} P(x)\, dx = 1$$

With a continuous distribution function and a continuous variable $M(x)$, the summation in Eq. (C.2) is replaced by an integral, so that the mean value of the measured quantity M becomes

$$\bar{M} = \frac{\int M(x)\, dN}{\int dN} = \int_{\text{all } x} M(x)P(x)\, dx \qquad (C.4)$$

These concepts can easily be extended to a multidimensional distribution function $P(x_1, x_2, \ldots)\, dx_1\, dx_2 \cdots$, which represents the probability that *each* variable will be within a range x_1 to $x_1 + dx_1$, x_2 to $x_2 + dx_2$, etc. A less–restrictive specification—for example, the probability of finding a system with x_1 between x_1 to $x_1 + dx_1$ while all possible values of the other variables are included—is given by

$$P(x_1)\, dx_1 = dx_1 \int_{x_2} \int_{x_3} \cdots \int P(x_1, x_2, \ldots)\, dx_2\, dx_3 \ldots \qquad (C.5)$$

C.2. The Boltzmann Distribution

We now ask the following question: Given a large number of identical molecules (or atoms or ions) and knowing from quantum mechanics the allowed energy levels of each molecule, what is the most probable distribution of molecules among the various energy levels?

Each individual molecule has a number of energy levels $1, 2, \ldots, i, \ldots$, with energy $\epsilon_1, \epsilon_2, \ldots, \epsilon_i, \ldots$; it is not necessary at this point to specify the kind of energy. Each level has associated with it a weighting factor or degeneracy g_i, which simply means that there are g_i wave functions that satisfy the Schrödinger equation for energy ϵ_i. From the total collection of N molecules there are N_1 with energy $\epsilon_1, \ldots, N_i$ with energy ϵ_i, etc. Thus a specification of all the occupation numbers $N_1, N_2, \ldots, N_i, \ldots$, is a particular state of the ensemble of systems (molecules). Furthermore, the total energy of the collection is fixed at a value E. The assumption that each molecule is distinguishable from all others and that the number which can be placed in each level is unrestricted leads to Boltzmann or "classical" statistics.

The probability of any particular distribution N_1, N_2, etc., is equal to the number of ways of realizing that distribution. There are $N!$

permutations of all the N molecules, but since the permutation of the N_i molecules within a given level does not lead to a new distribution, the probability is

$$W' = \frac{N!}{N_1! N_2! \cdots N_j!} = \frac{N!}{\Pi_j N_j!} \tag{C.6}$$

We must also include an additional factor arising from the degeneracies. In each level, the number of different ways in which the N_j molecules can be distributed among the g_j quantum states is $g_j^{N_j}$. (With no restrictions on the number of particles per state this result follows, for there are g_j choices of state for each molecule.)

Finally, the number of ways a given distribution can arise is

$$W = \left(\frac{N!}{\Pi_j N_j!}\right)(\Pi_j g_j^{N_j}) = N! \Pi \frac{g_j^{N_j}}{N_j!} \tag{C.7}$$

Equation (C.7) can be written in the form

$$\ln W = \ln N! + \ln \Pi\left(\frac{g_j^{N_j}}{N_j!}\right)$$
$$= \ln N! + \Sigma(N_j \ln g_j - \ln N_j!) \tag{C.8}$$

The factorial can be expressed by the Stirling approximation

$$\ln N! = N \ln N - N$$

Thus

$$\ln W = (N \ln N - N) + \Sigma(N_j \ln g_j - N_j \ln N_j + N_j)$$
$$= N \ln N + \Sigma(N_j \ln g_j - N_j \ln N_j) \tag{C.9}$$

The maximum value of W can be found by differentiating this expression and equating the derivative to zero. Since N and each g_j are constants, the result is

$$d \ln W = \Sigma(\ln g_j \, dN_j - \ln N_j \, dN_j - dN_j) = 0 \tag{C.10}$$

Conservation of the total number of molecules and energy gives

$$N = \Sigma_j N_j$$
$$dN = 0 = \Sigma dN_j \tag{C.11}$$

and

$$E = \Sigma_j N_j \epsilon_j$$
$$dE = 0 = \Sigma \epsilon_j \, dN_j \tag{C.12}$$

Hence

$$d \ln W = \Sigma \ln\left(\frac{g_j}{N_j}\right) dN_j = 0 \tag{C.13}$$

These three equations can be solved by the method of Lagrange multipliers. Equations (C.11) and (C.12) are multiplied by arbitrary constants α and β, and added to (C.13), to give

$$\Sigma\alpha\, dN_j + \Sigma\beta\epsilon_j\, dN_j + \Sigma \ln\left(\frac{g_j}{N_j}\right) dN_j = 0$$

$$\Sigma\left[\alpha + \beta\epsilon_j + \ln\left(\frac{g_j}{N_j}\right)\right] dN_j = 0 \qquad\qquad (C.14)$$

For this expression to be valid for any arbitrary choices of dN_1, dN_2, etc., each coefficient must be identically zero. That is,

$$\alpha + \beta\epsilon_j + \ln\left(\frac{g_j}{N_j}\right) = 0 \qquad\qquad (C.15)$$

for each value of j.

This expression is more commonly written

$$N_j = g_j e^{-\alpha} e^{-\beta\epsilon_j} \qquad\qquad (C.16)$$

Furthermore,

$$\frac{N_j}{N} = \frac{e^{-\alpha} g_j e^{-\beta\epsilon_j}}{e^{-\alpha}\Sigma g_j e^{-\beta\epsilon_j}}$$

$$\frac{N_j}{N} = \frac{g_j e^{-\beta\epsilon_j}}{\Sigma g_j e^{-\beta\epsilon_j}} \qquad\qquad (C.17)$$

The parameter β must be evaluated empirically. The usual method is to compare the equation of state derived from Eq. (C.17) for an ideal gas with the ideal gas law, from which β is found to be $(kT)^{-1}$. Finally, we have the Boltzmann distribution law

$$\frac{N_j}{N} = \frac{g_j e^{-\epsilon_j/kT}}{\Sigma g_j e^{-\epsilon_j/kT}} \qquad\qquad (C.18)$$

The quantity $\exp\left(-\epsilon_j/kT\right)$ is often called a Boltzmann factor; it can range from zero to unity, and it represents the "availability" or accessibility of state j at temperature T.

C.3. Partition Functions

The denominator of Eq. (C.18) is of sufficiently general use to deserve a name; it is the *partition function*

$$Q = \Sigma_j g_j e^{-\epsilon_j/kT} \qquad\qquad (C.19)$$

The partition function is a dimensionless number, greater than or equal to unity, representing the number of states "available" to the molecule at temperature T.

It is usually possible to factor the molecular partition function. The energy of a molecule, to a high degree of accuracy, can be expressed as a sum of translational, rotational, vibrational, and electronic terms:

$$\epsilon = \epsilon_{\text{trans}} + \epsilon_{\text{rot}} + \epsilon_{\text{vib}} + \epsilon_{\text{el}} \tag{C.20}$$

This condition, together with the fact that the corresponding degeneracies (g_{trans}, etc.) are multiplicative, leads to

$$Q = Q_{\text{trans}} Q_{\text{rot}} Q_{\text{vib}} Q_{\text{el}} \tag{C.21}$$

The translational partition function can be obtained from Eq. (C.19) by substituting for the energy levels those obtained from the quantum mechanical treatment of a particle in a box. However, an instructive alternative procedure is outlined in the next section.

Without discussing the methods of deriving expressions for the other factors in Eq. (C.21) (see, for example, reference 1), we simply set out the results.

The rotational partition function is

$$Q_{\text{rot}} = 8\pi^2 \frac{IkT}{sh^2} \qquad \text{(linear molecule)}$$

$$Q_{\text{rot}} = \left(\frac{8\pi^2}{s}\right)(8\pi^3 I_A I_B I_C)^{1/2} \left(\frac{kT}{h^2}\right)^{3/2} \qquad \text{(nonlinear molecule)} \tag{C.22}$$

In these expressions, I is the single moment of inertia for a linear molecule

$$I = \Sigma m_i r_i^2$$

where m_i is the mass of each atom and r_i is the distance of the atom from the center of mass; I_A, I_B, and I_C are moments of inertia about the principal axes for a nonlinear molecule; and s is the symmetry number (1 for CO or HD, 2 for H_2O or H_2, 3 for NH_3, 12 for CH_4, etc.) and is the number of indistinguishable positions into which the molecule can be turned by simple rigid rotations. The assumption has been made in deriving Eq. (C.22) that rotational energy level spacings are much smaller than the mean thermal energy. This assumption is valid, at chemically interesting temperatures, for all molecules but those containing hydrogen; small quantum corrections may be needed for them.

A nonlinear molecule of r atoms has three translational degrees of freedom, three rotational degrees of freedom, and $3r - 6$ vibrational degrees of freedom. In a linear molecule, one degree of rotational freedom is transformed into an additional vibration. The vibrational energy of a polyatomic molecule can be expressed in terms of the frequencies of the $3r - 6$ (or $3r - 5$) particular vibrations called the normal modes. For each of these normal modes, the vibrational partition function can be expressed simply if the vibrational levels are equally spaced, as they would be for a quantitized harmonic oscillator. With the lowest vibrational level as the reference energy, the vibrational partition function is

$$Q_{vib} = (1 - e^{-u})^{-1} \qquad (C.23)$$

where $u = hv/kT$ and v is the vibrational frequency. When v is in the customary units of wave numbers, the constant h/k has the value 1.4387. Each of the $3r - 6$ or $3r - 5$ normal mode vibrations of a molecule contributes a factor like (C.23). If the reference energy is moved to the minimum of the potential energy curve, Eq. (C.23) becomes

$$Q_{vib} = e^{-u/2}(1 - e^{-u})^{-1} \qquad (C.24)$$

The classical limit of Q_{vib} is found when the vibrational frequency is very low or the temperature very high. Equation (C.24) becomes

$$(Q_{vib})_{classical} = \frac{1}{u}$$

and we define a ratio

$$\Gamma_{vib} = \frac{Q_{vib}}{(Q_{vib})_{classical}}$$

or

$$\Gamma_{vib} = \frac{ue^{-u/2}}{1 - e^{-u}} = \frac{u/2}{\sinh(u/2)} \qquad (C.25)$$

If u is small, this can be expanded in a series

$$\Gamma_{vib} = \left[1 + \left(\frac{u^2}{24}\right) + \left(\frac{u^4}{1920}\right) + \cdots \right]^{-1} \qquad (C.26)$$

The electronic partition function must be evaluated term by term from the defining expression (C.19), where g_i and ϵ_i become the multiplicity and energy of electronic states. In most cases, the first excited level is so far above the ground state that Q_{el} is merely the ground-state multiplicity. Furthermore, rotational and vibrational properties of excited states are generally very different from those of the ground state. It is thus reasonable to consider the excited state as if it were an entirely different molecule, with its own origin for vibrational and rotational energy levels.

C.4. The Classical Partition Function: The Velocity Distribution

The classical form of Boltzmann's law requires that the molecular energy be treated as a continuous and not a quantized variable. For each molecule of r atoms, the total energy is given by

$$\epsilon = \frac{1}{2} \sum_{i=1}^{3r} m_i \dot{q}_i^2 + V(q_1, \ldots, q_{3r})$$

$$= \frac{1}{2} \sum \frac{p_i^2}{m_i} + V(q_1, \ldots, q_{3r}) \tag{C.27}$$

where the q's and p's are appropriate coordinates and the corresponding momenta, and V is the potential energy. To define the state of such a molecule, not only must the total energy be specified but so must a value of each coordinate and momentum. This corresponds to specifying a point at a position $q_1, q_2, \ldots, q_{3r}, p_1, p_2, \ldots, p_{3r}$, in a $6r$ — dimensional hyperspace called "phase space." However, if we want a smooth transition between the classical picture and the quantum mechanical description, the Uncertainty Principle must be invoked. As applied here, this principle permits us to define the state of the molecule only to the extent that the representative point in phase space is within the volume element $dq_1 \cdots dp_{3r} = h^{3r}$. With this restriction, the classical partition function is

$$Q = h^{-3r} \int_{-\infty}^{+\infty} \cdots \int_{-\infty}^{+\infty} e^{-\epsilon/kT} dq_1 \cdots dp_{3r} \tag{C.28}$$

Thus the summation of Eq. (C.19) has been replaced by a multiple integral over all possible values of coordinates and momenta of the individual atoms, and the weighting factor for each degree of freedom has been replaced by a term $dq\,dp/h$. The Boltzmann law becomes

$$\frac{dN}{N} = e^{-\epsilon/kT} \frac{dq_1 \cdots dp_{3r}}{h^{3r}Q} \tag{C.29}$$

where dN/N is the fraction of molecules with coordinate q_1 in the range q_1 to $q_1 + dq_1$, q_2 in the range q_2 to $q_2 + dq_2$, etc. Equation (C.29) is a specific example of the continuous distribution function defined in Section C.1.

The preceding equation can be used to obtain an expression for the velocity distribution in an ideal gas at thermal equilibrium. Equations (C.20) and (C.21) make it possible to separate the three translational degrees of freedom of each gas molecule from its internal degrees of freedom. In view of this fact, and since there are no external forces acting on a molecule in an ideal gas, $V(q_1, \ldots, q_{3r})$ in Eq. (C.27) disappears.

The kinetic energy term is simply that for a structureless molecule of mass $M = \sum_i m_i$—that is

$$\epsilon = \tfrac{1}{2}M(\dot{x}^2 + \dot{y}^2 + \dot{z}^2)$$

where $\dot{x}$, $\dot{y}$, and $\dot{z}$ are velocities with respect to an arbitrary Cartesian coordinate system. The Boltzmann equation is now:

$$\frac{dN(x, y, z, \dot{x}, \dot{y}, \dot{z})}{N} = (h^3 Q)^{-1} \exp\left[\frac{-M(\dot{x}^2 + \dot{y}^2 + \dot{z}^2)}{2kT}\right]$$

$$\times (dx\ dy\ dz)(M^3 d\dot{x}\ d\dot{y}\ d\dot{z}) \qquad (C.30)$$

We can immediately obtain the translational partition function by integrating over the coordinates x, y, and z. Limits are provided by the dimensions of the container, assumed to be a box with sides of length a, b, and c, and volume V. The partition function is

$$Q_{trans} = h^{-3}\left[\int_0^a \int_{-\infty}^{\infty} \exp\left(\frac{-p_x^2}{2MkT}\right) dx\ dp_x\right]\left[\int_0^b \int_{-\infty}^{\infty} \exp\left(\frac{-p_y^2}{2MkT}\right) dy\ dp_y\right]$$

$$\times \left[\int_0^c \int_{-\infty}^{\infty} \exp\left(\frac{-p_z^2}{2MkT}\right) dz\ dp_z\right]$$

$$= h^{-3}abc\left[\int_{-\infty}^{\infty} \exp\left(\frac{-p^2}{2MkT}\right)\right]^3$$

The remaining definite integral is a standard form, and the partition function is finally

$$Q_{trans} = \left(\frac{2\pi MkT}{h^2}\right)^{3/2} V$$

or, per unit volume,

$$Q'_{trans} = \left(\frac{2\pi MkT}{h^2}\right)^{3/2} \qquad (C.31)$$

In considering the velocity distribution, it is convenient to change from cartesian coordinates to spherical coordinates r, θ, and ϕ such that

$$r = (x^2 + y^2 + z^2)^{1/2}$$
$$x = r \sin\theta \cos\phi$$
$$y = r \sin\theta \sin\phi$$
$$z = r \cos\theta,$$

and

$$dV = dx\ dy\ dz = r^2 \sin\theta d\theta d\phi\ dr$$

The corresponding relations among the velocities lead to

$$c^2 = \dot{x}^2 + \dot{y}^2 + \dot{z}^2$$

and

$$dp_x\ dp_y\ dp_z = M^3 c^2 \sin\theta d\theta d\phi\ dc$$

In terms of the new coordinate system, Eq. (C.30) becomes

$$\frac{dN(r, \theta, \phi, c)}{N} = (h^3 Q)^{-1} M^3 e^{-Mc^2/2kT}\ dVc^2 \sin\theta\ d\theta\ d\phi\ dc \qquad (C.32)$$

This expression can be integrated over the entire volume, and when

Eq. (C.31) for Q_{trans} is introduced, the result is

$$\frac{dN(\theta, \phi, c)}{N} = \frac{M^3}{(2\pi MkT)^{3/2}} e^{-Mc^2/2kT} c^2 dc \sin \theta \, d\theta \, d\phi \tag{C.33}$$

However, our immediate interest is in the distribution function $P(c) \, dc$, the fraction of molecules with speeds in the range c to $c + dc$, regardless of the directions in which they are moving. This factor is obtained by further integration over the angle variables to obtain the factor 4π—that is, the solid angle subtended by a sphere. Finally,

$$\frac{dN(c)}{N} = P(c) \, dc = 4\pi \left(\frac{M}{2\pi kT}\right)^{3/2} e^{-Mc^2/2kT} c^2 \, dc \tag{C.34}$$

The mean speed can be obtained in the usual way from Eq. (C.4):

$$\bar{c} = \int_0^\infty cP(c) \, dc$$

$$= 4\pi \left(\frac{M}{2\pi kT}\right)^{3/2} \int_0^\infty c^3 e^{-Mc^2/2kT} \, dc$$

The integral is a standard form, and one finds that

$$\bar{c} = \left(\frac{8kT}{\pi M}\right)^{1/2} \tag{C.35}$$

C.5. The Relation Between Partition Functions and Equilibrium Constants

A system at chemical equilibrium, such as

$$A + B \rightleftharpoons C$$

is more complicated than that discussed in Section C.2, for it contains different molecular species. However, an analogous calculation of the most probable distribution can be made. This calculation leads to the following expression for the equilibrium constant (in units of cm^3 molecule^{-1}):

$$K_{eq} = \frac{N_C}{N_A N_B} = \frac{Q'_C}{Q_A Q_B} \tag{C.36}$$

The partition function for C is primed as it is computed relative to the energy origins of A and B. If the partition function of C is specified with zero of its energy scale at the electronic energy minimum of C, $-D_e(C)$, then the energy difference

$$\Delta\epsilon = D_e(A) + D_e(B) - D_e(C)$$

can be removed from Q_C' as a Boltzmann factor, giving

$$K_{eq} = \frac{Q_C}{Q_A Q_B} e^{-\Delta\epsilon/kT} \tag{C.37}$$

which is a more useful form than Eq. (C.36).

The concentration of one particular state can be found by combining the Boltzmann law (Eq. C.29) with (C.36), giving

$$\frac{dN_C}{N_A N_B} = \frac{e^{-\Delta\epsilon/kT} dq_1 \cdots dp_{3r}}{h^{3r} Q_A Q_B} \tag{C.38}$$

Reference

1. For example, W. J. Moore, *Physical Chemistry*, 3rd ed., Prentice-Hall, Inc., Englewood Cliffs, N.J., 1962, Chapter 15.

Index